# Universitext

# Universitext

*Universitext* is a series of textbooks that presents material from a wide variety of mathematical disciplines at master's level and beyond. The books, often well class-tested by their author, may have an informal, personal even experimental approach to their subject matter. Some of the most successful and established books in the series have evolved through several editions, always following the evolution of teaching curricula, into very polished texts.

Thus as research topics trickle down into graduate-level teaching, first textbooks written for new, cutting-edge courses may make their way into *Universitext*.

More information about this series at http://www.springer.com/series/223

Sergei Ovchinnikov

# Functional Analysis

## An Introductory Course

 Springer

Sergei Ovchinnikov
Department of Mathematics
San Francisco State University
San Francisco, CA
USA

ISSN 0172-5939       ISSN 2191-6675   (electronic)
Universitext
ISBN 978-3-319-91511-1      ISBN 978-3-319-91512-8   (eBook)
https://doi.org/10.1007/978-3-319-91512-8

Library of Congress Control Number: 2018941226

Mathematics Subject Classification (2010): 46B25, 46C05, 46B15, 46B45

This Springer imprint is published by the registered company Springer Nature Switzerland AG
The registered company address is: Gewerbestrasse 11, 6330 Cham, Switzerland

For Galina (of course)

# Preface

There are many excellent textbooks on functional analysis. However, only few can be regarded as introductory or elementary. In writing this book, my goal was to produce a "really" introductory (and short) text on this subject.

This book originated from the notes that I kept while teaching the graduate course *Introductory Functional Analysis* at San Francisco State University (SFSU). The Mathematics Department of SFSU offers the Master of Arts in Mathematics degree to its students. Many students who received an MA degree from this department do not pursue a higher degree. When teaching this course I wanted my students to get a better feel for analysis in general, appreciate its importance, and be ready to learn more about it, should the need arise. Some of these students had gaps in their knowledge of basic analysis and other branches of mathematics. For this reason, in addition to functional analysis topics, the book contains materials that any undergraduate/graduate student of mathematics should know. These materials are subjects of the first and second chapters.

The first chapter presents selected topics of general mathematical interest that I felt are needed to master the material presented in the book. First, it includes a brief overview of convex functions with applications to classical inequalities, that will be used in the succeeding chapters. Second, binary relations of equivalence and order and their main properties are described. The chapter ends with Zorn's lemma, which is essential in the establishing some advanced results later in the book.

Most spaces in functional analysis are topological (often metric) vector spaces. This is why natural prerequisites for a course in functional analysis include courses in linear algebra and real analysis. I do not include a preliminary overview of linear algebra. However, I felt that it is important to review basics of metric and topological spaces to fill the gaps between what the reader probably learned some time ago and what is necessary to master topics covered in the book. This review is the subject of the second chapter. In addition to standard topics on metric spaces covered in real analysis, I prove a "light" version of the Baire Category Theorem in Chapter 2. A brief discussion of this theorem is found at the end of the chapter.

Metric and topological vector spaces are introduced in the first section of Chapter 3. The rest of this chapter is devoted to "special spaces" that serve as illustrations to many general theorems in the book. These spaces are finite-dimensional spaces $\ell_p^n$, sequence spaces $\ell_p$, $c$, $c_0$, and $\mathbf{s}$, and spaces of continuous functions and functions of bounded variation. Because I wanted to keep this book on an "introductory" level, I did not include in this chapter spaces of Lebesgue measurable functions despite their great role in functional analysis. The main thrust of this chapter is on establishing properties such as completeness and separability of the spaces under consideration.

Normed spaces, bounded linear operators on these spaces, and their properties are subjects of Chapter 4. The concept of a Schauder basis (cf. Section 4.1) illustrates the advantage of blending the metric and linear structures of normed spaces allowing for formation of convergent series. The central result of Section 4.2 is equivalence of continuity and boundedness properties of linear operators, thus taking an advantage of the vector space structure in establishing analytical properties of continuous linear operators. Fundamental topological properties of finite-dimensional normed spaces are discussed in Sections 4.3 and 4.4.

A separate chapter in the book is devoted to linear functionals. First, in Section 5.1, it is demonstrated that many normed spaces have considerably "large" dual spaces, that is, spaces of continuous linear functionals on the underlying space. In Sections 5.2 and 5.3 various results known as the Hahn–Banach Theorem are established. This theorem plays a fundamental role in functional analysis. In particular, the Hahn–Banach Theorem shows that there are "enough" continuous linear functionals on a normed space to motivate the study of the dual space. As an application of this theorem, all bounded linear functionals on the space $C[a, b]$ are described in Section 5.4. In conclusion of Chapter 5, reflexive spaces are introduced and their elementary properties established.

Three "pillars" of functional analysis—the Uniform Boundedness, Open Mapping, and Closed Graph Theorems—are covered in Chapter 6. (The forth "pillar" is the Hahn–Banach Theorem (cf. Chapter 5)). The distinguished role of the completeness property in establishing these fundamental results is illustrated by a number of examples and counterexamples in Section 6.4.

It is hard to imagine a textbook on functional analysis that does not include topics on Hilbert spaces. Chapter 7 (that constitutes about 20% of the entire book) is an in introduction to the theory of Hilbert spaces. It begins with a short coverage of inner product spaces in Section 7.1. The next two sections introduce and discuss the orthogonality property in inner product spaces and orthonormal families in Hilbert spaces. In my opinion, the concept of summable families of vectors in normed spaces is elementary enough to be included into an introductory course in functional analysis. This notion is used in Section 7.3 to define Hilbert bases and establish convenient representation of vectors in Hilbert spaces. In the same section, the Gram–Schmidt process is used to construct orthonormal bases in separable Hilbert spaces. Effective tools for studying linear functionals and operators on Hilbert spaces are the Riesz Representation theorem and sesquilinear forms that are covered in Sections 7.4 and 7.5, respectively. The concept of Hilbert-adjoint operator (cf. Section 7.6) is used to introduce the standard classes of normal, self-adjoint, and unitary operators in Section 7.7. Projection operators are covered in the last section of this chapter.

To enhance the presentation of Hilbert spaces, more advanced topics on summable families of vectors in a normed space are included in the Appendix A, where Hilbert spaces $\ell_2(J)$ are described for arbitrary set $J$.

As I mentioned at the beginning of this Preface, I wanted to write an introductory text on functional analysis. In fact, I wanted to keep the book on "elementary" level. For this reason, the book does not rely on the Lebesgue theory and does not include the Spectral theorem. However, some topics that are usually considered as not completely elementary are included in the book. For instance, Zorn's lemma, Baire's Category theorem and summability of families of vectors in a normed space are included in the book.

In my opinion, the most effective way of learning mathematics is by "doing it". There are 240 exercises in the book. Most exercises are "proof" problems, that is, the reader is invited to prove a statement in the exercise. Every chapter has its own set of exercises.

Throughout the book the symbol $\mathbf{F}$ denotes either the real field, $\mathbf{R}$, or the complex field, $\mathbf{C}$. Symbols $\mathbf{N}$, $\mathbf{Z}$, and $\mathbf{Q}$ denote the sets of natural numbers $\{1, 2, \ldots\}$, integers, and rational numbers, respectively. The symbol $\subseteq$ is always used for the set inclusion relation, proper or not.

I wish to thank Sheldon Axler for his endorsement of this project and an anonymous referee for reading the manuscript carefully and suggesting mathematical and stylistic corrections. My special thanks go to Eric Hayashi for numerous comments and recommendations which materially improved the original draft of the book.

Berkeley, California                                    Sergei Ovchinnikov
March 2018

# Contents

# Chapter 1
# Preliminaries

In this chapter, we briefly cover topics that are ubiquitous in mathematics and are included in this book for the reader's convenience.

Sections 1.1 and 1.2 introduce basic inequalities that are used later in the book. Although spaces of Lebesgue integrable functions are not covered in the book, some integral inequalities are covered in Section 1.3.

Equivalence and order relations and Zorn's lemma are the topics of Sections 1.4 and 1.5.

## 1.1 Convex Functions

In this section, $I$ is an open interval on the real line $\mathbf{R}$ that may be bounded or unbounded. A function $f : I \to \mathbf{R}$ is said to be *convex* if

$$f[\lambda x + (1 - \lambda)y] \leq \lambda f(x) + (1 - \lambda)f(y), \qquad (1.1)$$

for all $x, y \in I$ and $\lambda \in [0, 1]$. It is called *concave* if the opposite inequality

$$f[\lambda x + (1 - \lambda)y] \geq \lambda f(x) + (1 - \lambda)f(y) \qquad (1.2)$$

holds for all $x, y \in I$ and $\lambda \in [0, 1]$. Note that the numbers $\lambda x + (1 - \lambda)y$ in (1.1) and (1.2) belong to the domain $I$ of the function $f$ (cf. Exercise 1.1).

Geometrically (cf. Fig. 1.1), inequality (1.1) means that each point $R$ on the graph of the function $f$ that lies between two other points $P$ and $Q$ on the graph lies on or below the chord $PQ$ (cf. Exercise 1.2). In the case of inequality (1.2), $R$ lies on or above the chord $PQ$.

It is clear that a function $f$ is concave if and only if the function $-f$ is convex.

The following theorem is instrumental in proving many inequalities that appear later in this chapter.

© Springer International Publishing AG, part of Springer Nature 2018
S. Ovchinnikov, *Functional Analysis*, Universitext,
https://doi.org/10.1007/978-3-319-91512-8_1

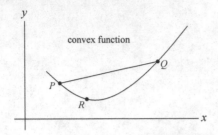 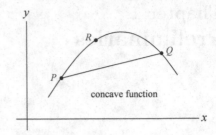

**Fig. 1.1** Convex and concave functions.

**Theorem 1.1.** *A function $f : I \to \mathbf{R}$ is convex if and only if*

$$f\left(\sum_{k=1}^{n} \lambda_k x_k\right) \le \sum_{k=1}^{n} \lambda_k f(x_k), \tag{1.3}$$

*for all $x_1, \ldots, x_n \in I$ and $0 \le \lambda_k \le 1$ $(1 \le k \le n)$ such that $\sum_{k=1}^{n} \lambda_k = 1$.*

Note that the number $\sum_{k=1}^{n} \lambda_k x_k$ belongs to $I$ (cf. Exercise 1.1), so the left-hand side in (1.3) is well defined.

*Proof.* (Necessity.) We prove by induction that (1.3) holds for every convex function $f$.

The base case, $n = 1$, holds trivially.

Suppose that (1.3) holds for some $m \ge 1$ and let

$$x_1, \ldots, x_m, x_{m+1} \in I, \quad 0 \le \lambda_k \le 1\,(1 \le k \le m+1), \quad \sum_{k=1}^{m+1} \lambda_k = 1.$$

We may assume that $\lambda_{m+1} \ne 1$. (Otherwise, the induction step is trivial.) For $\lambda = \sum_{k=1}^{m} \lambda_k$, we have $\lambda_{m+1} = 1 - \lambda$, and then by (1.1) and the induction hypothesis,

$$
\begin{aligned}
f\left(\sum_{k=1}^{m+1} \lambda_k x_k\right) &= f\left(\lambda \sum_{k=1}^{m} \frac{\lambda_k}{\lambda} x_k + (1-\lambda) x_{m+1}\right) \\
&\le \lambda f\left(\sum_{k=1}^{m} \frac{\lambda_k}{\lambda} x_k\right) + \lambda_{m+1} f(x_{m+1}) \\
&\le \lambda \sum_{k=1}^{m} \frac{\lambda_k}{\lambda} f(x_k) + \lambda_{m+1} f(x_{m+1}) \\
&= \sum_{k=1}^{m+1} \lambda_k f(x_k).
\end{aligned}
$$

The result follows by induction.

(Sufficiency.) This follows immediately from (1.3) for $n = 2$. $\qquad\square$

Similarly, a function $f : I \to \mathbf{R}$ is concave if and only if

$$f\left(\sum_{k=1}^{n} \lambda_k x_k\right) \geq \sum_{k=1}^{n} \lambda_k f(x_k), \tag{1.4}$$

for all $x_1, \ldots, x_n \in I$ and $0 \leq \lambda_k \leq 1$ $(1 \leq k \leq n)$ such that $\sum_{k=1}^{n} \lambda_k = 1$.

The next theorem is a useful characterization of smooth convex functions. The proof is left to the reader (cf. Exercise 1.3).

**Theorem 1.2.** *If $f$ is a differentiable function on $I$, then it is convex if and only if $f'$ is increasing on $I$. Similarly, $f$ is concave if and only if $f'$ is decreasing on $I$.*

*Example 1.1.* (Cf. Fig. 1.2.) Let $f(x) = x^p$, $p \geq 1$, be a power function on $(0, \infty)$. Because $f'(x) = px^{p-1}$ is increasing on $(0, \infty)$, $f$ is a convex function.

On the other hand, the natural logarithm function $f(x) = \ln x$ on $(0, \infty)$ is concave, because $f'(x) = 1/x$ is decreasing on $(0, \infty)$.

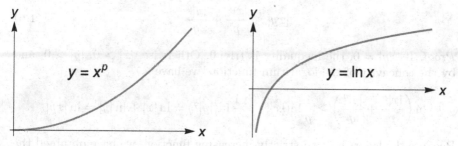

**Fig. 1.2** The power and logarithmic functions.

## 1.2 Inequalities

In this section, unless otherwise specified, all variables assume values in the field of complex numbers $\mathbf{C}$. Recall that $|z|$ stands for the absolute value (modulus) of a complex number $z$, $\bar{z}$ stands for the complex conjugate of $z$, $\operatorname{Re}(z)$ and $\operatorname{Im}(z)$ denote the real and imaginary parts of $z$, respectively, and

$$|z|^2 = z \cdot \bar{z}, \quad 2\operatorname{Re}(z) = z + \bar{z}, \quad 2i\operatorname{Im}(z) = z - \bar{z},$$

for every $z \in \mathbf{C}$.

We begin by proving the *triangle inequality* for complex numbers:

$$|x + y| \leq |x| + |y|, \qquad \text{for all } x, y \in \mathbf{C}. \tag{1.5}$$

Indeed,

$$|x + y|^2 = (x + y)(\overline{x + y}) = (x + y)(\bar{x} + \bar{y}) = x\bar{x} + x\bar{y} + y\bar{x} + y\bar{y}$$
$$= |x|^2 + x\bar{y} + \overline{x\bar{y}} + |y|^2 = |x|^2 + 2\mathrm{Re}(x\bar{y}) + |y|^2$$
$$\leq |x|^2 + 2|x||y| + |y|^2 = (|x| + |y|)^2,$$

because $\mathrm{Re}(z) \leq |z|$ for every complex number $z$. From this, (1.5) follows immediately. Of course, the triangle inequality also holds for real numbers.

For a real number $p > 1$ we define $q$ by the equation

$$\frac{1}{p} + \frac{1}{q} = 1.$$

Clearly, there is a unique real number $q > 1$ satisfying this equation. The numbers $p$ and $q$ are called *conjugate exponents*.

**Theorem 1.3. (Young's inequality.)** *If $p$ and $q$ are conjugate exponents, then for all complex numbers $x$ and $y$,*

$$|xy| \leq \frac{|x|^p}{p} + \frac{|y|^q}{q}. \tag{1.6}$$

*Proof.* If $|xy| = 0$, the inequality is trivial. Otherwise, $|x| > 0$, $|y| > 0$, and by the concavity of the logarithm function, we have

$$\ln\left(\frac{|x|^p}{p} + \frac{|y|^q}{q}\right) \geq \frac{1}{p}\ln(|x|^p) + \frac{1}{q}\ln(|y|^q) = \ln|x| + \ln|y| = \ln|xy|.$$

Because the logarithm is a strictly increasing function, we have obtained the desired result.  □

**Theorem 1.4. (Hölder's inequality.)** *If $p$ and $q$ are conjugate exponents, then*

$$\sum_{k=1}^{n} |x_k y_k| \leq \left(\sum_{k=1}^{n} |x_k|^p\right)^{1/p} \left(\sum_{k=1}^{n} |y_k|^q\right)^{1/q} \tag{1.7}$$

*for arbitrary complex numbers $x_1, \ldots, x_n$ and $y_1, \ldots, y_n$.*

*Proof.* We may assume that at least one of the $|x_k|$'s and one of the $|y_k|$'s are positive. (Otherwise, both sides in (1.7) are zero and we are done.) Then the real numbers

$$u = \left(\sum_{k=1}^{n} |x_k|^p\right)^{1/p} \quad \text{and} \quad v = \left(\sum_{k=1}^{n} |y_k|^q\right)^{1/q}$$

are positive. We apply Young's inequality (1.6) to $|x_k|/u$ and $|y_k|/v$ to obtain

$$\frac{|x_k|}{u} \cdot \frac{|y_k|}{v} \leq \frac{1}{p}\left(\frac{|x_k|}{u}\right)^p + \frac{1}{q}\left(\frac{|y_k|}{u}\right)^q, \qquad 1 \leq k \leq n,$$

or, after summing from $k = 1$ to $k = n$,

$$\frac{\sum_{k=1}^{n}|x_k y_k|}{uv} \leq \frac{1}{p} \cdot \frac{\sum_{k=1}^{n}|x_k|^p}{u^p} + \frac{1}{q} \cdot \frac{\sum_{k=1}^{n}|y_k^q|}{v^q} = \frac{1}{p} + \frac{1}{q} = 1,$$

which is equivalent to (1.7). $\qquad\qquad\qquad\qquad\qquad\qquad\square$

**Theorem 1.5. (Cauchy–Schwarz inequality.)** *For arbitrary complex numbers $x_1, \ldots, x_n$ and $y_1, \ldots, y_n$,*

$$\sum_{k=1}^{n}|x_k y_k| \leq \sqrt{\sum_{k=1}^{n}|x_k|^2}\sqrt{\sum_{k=1}^{n}|y_k|^2}. \tag{1.8}$$

Inequality (1.8) follows from Hölder's inequality by setting $p = q = 2$. We present an elementary proof that does not rest on the theory of convex functions.

*Proof.* We have (cf. the proof of (1.5))

$$0 \leq \sum_{k=1}^{n}|tx_k + y_k|^2 \leq t^2\sum_{k=1}^{n}|x_k|^2 + 2t\sum_{k=1}^{n}|x_k y_k| + \sum_{k=1}^{n}|y_k|^2,$$

for all real numbers $t$. Hence the discriminant of the quadratic trinomial on the right-hand side must be nonpositive:

$$4\left(\sum_{k=1}^{n}|x_n y_n|\right)^2 - 4\left(\sum_{k=1}^{n}|x_k|^2\right)\left(\sum_{k=1}^{n}|y_k|^2\right) \leq 0,$$

which is equivalent to (1.8). $\qquad\qquad\qquad\qquad\qquad\qquad\square$

The next theorem establishes an inequality that plays a distinguished role in analysis.

**Theorem 1.6. (Minkowski's inequality.)** *For arbitrary complex numbers $x_1, \ldots, x_n, y_1, \ldots, y_n$ and a real number $p \geq 1$,*

$$\left[\sum_{k=1}^{n}|x_k + y_k|^p\right]^{1/p} \leq \left[\sum_{k=1}^{n}|x_k|^p\right]^{1/p} + \left[\sum_{k=1}^{n}|y_k|^p\right]^{1/p}. \tag{1.9}$$

*Proof.* We may assume that both real numbers

$$u = \left( \sum_{k=1}^{n} |x_k|^p \right)^{1/p} \quad \text{and} \quad v = \left( \sum_{k=1}^{n} |y_k|^p \right)^{1/p}$$

are positive. By the triangle inequality, we have

$$|x_k + y_k|^p \le (|x_k| + |y_k|)^p = \left( u \frac{|x_k|}{u} + v \frac{|y_k|}{v} \right)^p$$

$$= \left[ (u+v) \left( \frac{u}{u+v} \frac{|x_k|}{u} + \frac{v}{u+v} \frac{|y_k|}{v} \right) \right]^p$$

$$= (u+v)^p \left( \frac{u}{u+v} \frac{|x_k|}{u} + \frac{v}{u+v} \frac{|y_k|}{v} \right)^p.$$

Because

$$\frac{u}{u+v} + \frac{v}{u+v} = 1$$

and the power function $x^p$ is convex for $p \ge 1$, we have

$$\left( \frac{u}{u+v} \frac{|x_k|}{u} + \frac{v}{u+v} \frac{|y_k|}{v} \right)^p \le \frac{u}{u+v} \frac{|x_k|^p}{u^p} + \frac{v}{u+v} \frac{|y_k|^p}{v^p}.$$

Hence

$$|x_k + y_k|^p \le (u+v)^p \left( \frac{u}{u+v} \frac{|x_k|^p}{u^p} + \frac{v}{u+v} \frac{|y_k|^p}{v^p} \right).$$

By summing both sides of the above inequality, we obtain

$$\sum_{k=1}^{n} |x_k + y_k|^p \le (u+v)^p \left( \frac{u}{u+v} \frac{\sum_{k=1}^{n} |x_k|^p}{u^p} + \frac{v}{u+v} \frac{\sum_{k=1}^{n} |y_k|^p}{v^p} \right)$$

$$= (u+v)^p \left( \frac{u}{u+v} + \frac{v}{u+v} \right)$$

$$= (u+v)^p = \left[ \left( \sum_{k=1}^{n} |x_k|^p \right)^{1/p} + \left( \sum_{k=1}^{n} |y_k|^p \right)^{1/p} \right]^p,$$

which is equivalent to (1.9). $\qquad\qquad\square$

Hölder's inequality, the Cauchy–Schwarz inequality, and Minkowski's inequality also hold for infinite sequences of complex numbers.

**Theorem 1.7. (Hölder's inequality.)** *Let $p$ and $q$ be conjugate exponents. If the series $\sum_{k=1}^{\infty} |x_k|^p$ and $\sum_{k=1}^{\infty} |y_k|^q$ are convergent, then the series $\sum_{k=1}^{\infty} |x_k y_k|$ converges, and*

$$\sum_{k=1}^{\infty} |x_k y_k| \le \left( \sum_{k=1}^{\infty} |x_k|^p \right)^{1/p} \left( \sum_{k=1}^{\infty} |y_k|^q \right)^{1/q}. \tag{1.10}$$

**Theorem 1.8. (Cauchy–Schwarz inequality.)** *If the series $\sum_{k=1}^{\infty} |x_k|^2$ and $\sum_{k=1}^{\infty} |y_k|^2$ are convergent, then the series $\sum_{k=1}^{\infty} |x_k y_k|$ converges and*

$$\sum_{k=1}^{\infty} |x_k y_k| \leq \sqrt{\sum_{k=1}^{\infty} |x_k|^2} \sqrt{\sum_{k=1}^{\infty} |y_k|^2}. \tag{1.11}$$

**Theorem 1.9. (Minkowski's inequality.)** *Suppose that the series $\sum_{k=1}^{\infty} |x_k|^p$ and $\sum_{k=1}^{\infty} |y_k|^p$ converge for a real number $p \geq 1$. Then the series*

$$\sum_{k=1}^{\infty} |x_k + y_k|^p$$

*converges, and*

$$\left[ \sum_{k=1}^{\infty} |x_k + y_k|^p \right]^{1/p} \leq \left[ \sum_{k=1}^{\infty} |x_k|^p \right]^{1/p} + \left[ \sum_{k=1}^{\infty} |y_k|^p \right]^{1/p}. \tag{1.12}$$

Proofs of these theorems are straightforward and left as an exercise (cf. Exercise 1.7).

If $0 < p < 1$, then we have to invert the inequality sign in Minkowski's inequality (cf. Exercise 1.9). However, in this case, we have the following analogue of the inequality in (1.12).

**Theorem 1.10.** *Suppose that the series $\sum_{k=1}^{\infty} |x_k|^p$ and $\sum_{k=1}^{\infty} |y_k|^p$ converge for a real number $0 < p < 1$. Then the series*

$$\sum_{k=1}^{\infty} |x_k + y_k|^p$$

*converges, and*

$$\sum_{k=1}^{\infty} |x_k + y_k|^p \leq \sum_{k=1}^{\infty} |x_k|^p + \sum_{k=1}^{\infty} |y_k|^p. \tag{1.13}$$

The claim of the theorem follows immediately from the following lemma.

**Lemma 1.1.** *Let $p$ be a real number such that $0 < p < 1$. Then for all complex numbers $x$ and $y$,*

$$|x + y|^p \leq |x|^p + |y|^p.$$

*Proof.* Let $f$ be a real function of a real variable $t$ defined by

$$f(t) = t^p - (t + 1)^p + 1, \qquad \text{for } t \geq 0.$$

Because $f(0) = 0$ and

$$f'(t) = pt^{p-1} - p(t+1)^{p-1} = p\left(\frac{1}{t^{1-p}} - \frac{1}{(t+1)^{1-p}}\right) > 0, \qquad \text{for } t > 0,$$

we have $f(t) \geq 0$ for $t \geq 0$, so

$$(t+1)^p \leq t^p + 1, \quad \text{for } t \geq 0.$$

Clearly, we may assume that $y \neq 0$ in the lemma. Then

$$|x+y|^p \leq (|x|+|y|)^p = |y|^p\left(\left|\frac{x}{y}\right|+1\right)^p \leq |y|^p\left(\left|\frac{x}{y}\right|^p+1\right) = |x|^p + |y|^p,$$

as desired.                                                                          □

We conclude this section by establishing another useful inequality.

**Theorem 1.11.** *For all complex numbers $x$ and $y$,*

$$\frac{|x+y|}{1+|x+y|} \leq \frac{|x|}{1+|x|} + \frac{|y|}{1+|y|}. \tag{1.14}$$

*Proof.* The function

$$f(t) = \frac{t}{1+t}$$

of a real variable $t$ is increasing over $[0,\infty)$, because

$$f'(t) = \frac{1}{(1+t)^2} > 0, \qquad \text{for } t \geq 0.$$

Then, because $|x+y| \leq |x|+|y|$ (cf. (1.5)), we have

$$\frac{|x+y|}{1+|x+y|} \leq \frac{|x|+|y|}{1+|x|+|y|} = \frac{|x|}{1+|x|+|y|} + \frac{|y|}{1+|x|+|y|} \leq \frac{|x|}{1+|x|} + \frac{|y|}{1+|y|},$$

and the result follows.                                                              □

## 1.3 Integral Inequalities

In this section, the functions $x$ and $y$ are Lebesgue measurable functions on the unit interval.

The proofs of the theorems in this section are very similar to the proofs of the respective theorems in the previous section and are therefore omitted (cf. Exercise 1.12).

**Theorem 1.12. (Hölder's inequality.)** *If $p$ and $q$ are conjugate exponents and $\int_0^1 |x|^p < \infty$, $\int_0^1 |y|^q < \infty$, then $\int_0^1 |xy| < \infty$ and*

$$\int_0^1 |xy| \leq \left( \int_0^1 |x|^p \right)^{1/p} \left( \int_0^1 |y|^q \right)^{1/q}. \tag{1.15}$$

By setting $p = q = 2$ in Hölder's integral inequality (1.15), we obtain the integral version of the Cauchy–Schwarz inequality (1.11).

**Theorem 1.13. (Cauchy–Schwarz inequality.)** *Suppose that*

$$\int_0^1 |x|^2 < \infty \quad and \quad \int_0^1 |y|^2 < \infty.$$

*Then $\int_0^1 |xy| < \infty$ and*

$$\int_0^1 |xy| \leq \sqrt{\int_0^1 |x|^2} \sqrt{\int_0^1 |y|^2}. \tag{1.16}$$

**Theorem 1.14. (Minkowski's inequality.)** *If $p \geq 1$ and $\int_0^1 |x|^p < \infty$, $\int_0^1 |y|^p < \infty$, then*

$$\left( \int_0^1 |x+y|^p \right)^{1/p} \leq \left( \int_0^1 |x|^p \right)^{1/p} + \left( \int_0^1 |y|^p \right)^{1/p}. \tag{1.17}$$

The result of the following theorem follows immediately from Lemma 1.1.

**Theorem 1.15.** *Suppose that $\int_0^1 |x|^p < \infty$ and $\int_0^1 |y|^p < \infty$ for a real number $0 < p < 1$. Then $\int_0^1 |x+y|^p < \infty$ and*

$$\int_0^1 |x+y|^p \leq \int_0^1 |x|^p + \int_0^1 |y|^p. \tag{1.18}$$

## 1.4 Equivalence Relations

By definition, a *binary relation* on a set $X$ is a subset of the Cartesian product $X \times X$. For a binary relation $R$ on $X$ we often write $x R y$ instead of $(x, y) \in R$.

Among many kinds of binary relations, the equivalence relations and partial orders arguably are the most common. We introduce the basics of equivalence relations below and cover partial orders in the next section. It is customary to use one symbol, $\sim$, for various equivalence relations. We follow this convention here.

An *equivalence relation* on a set $X$ is a reflexive, symmetric, and transitive binary relation on $X$, that is, a relation $\sim$ on $X$ satisfying the following conditions for all $x, y, z \in X$:

$$x \sim x \qquad\qquad\qquad\qquad reflexivity,$$
$$x \sim y \ \text{ implies } \ y \sim x \qquad\qquad symmetry,$$
$$x \sim y \ \text{ and } \ y \sim z \ \text{ implies } \ x \sim z \qquad transitivity.$$

If $\sim$ is an equivalence relation on $X$ and $x \in X$, then the set

$$[x] = \{y \in X : y \sim x\}$$

is called the *equivalence class* of $x$ with respect to $\sim$. The set of equivalence classes in $X$ with respect to $\sim$ is called the *quotient set* of $X$ with respect to $\sim$ and denoted by $X/\sim$. The mapping $x \mapsto [x]$ that assigns to each $x \in X$ its equivalence class $[x]$ in $X$ is called the *canonical map* of $X$ onto $X/\sim$.

Here are two "extreme" examples of equivalence relations on a set $X$.

*Example 1.2.* (1) The "complete" relation $X \times X$. Here $x \sim y$ for all $x, y \in X$, and it is easy to verify that $\sim$ is an equivalence relation. It is clear that there is only one equivalence class in $X$, namely the set $X$ itself. Hence the quotient set is a singleton.

(2) The equality relation $=$ on $X$ is evidently an equivalence relation. The equivalence class of $x \in X$ is the singleton $\{x\}$.

A nontrivial instance of an equivalence relation appears in a formal definition of rational numbers.

*Example 1.3.* A fraction $m/n$ can be viewed as an ordered pair of integers $(m, n)$ with $n \neq 0$. Two fractions $(m, n)$ and $(p, q)$ are said to be *equivalent* if $mq = np$:

$$(m, n) \sim (p, q) \qquad \text{if} \qquad mq = np.$$

It is not difficult to verify that $\sim$ is indeed an equivalence relation on the set $\mathbf{Z} \times (\mathbf{Z} \setminus \{0\})$. The equivalence classes of $\sim$ are called *rational numbers*.

**Theorem 1.16.** *Let $X$ be a set and $\sim$ an equivalence relation on $X$. The equivalence classes with respect to $\sim$ partition the set $X$, that is, every element of $X$ belongs to one and only one equivalence class.*

*Proof.* By the reflexivity property, $x \in [x]$. It remains to show that every two equivalence classes are either identical or disjoint. Suppose that $[x] \cap [y] \neq \varnothing$ and let $z \in [x] \cap [y]$. Then $z \sim x$ and $z \sim y$. By symmetry, $x \sim z$. For every $a \in [x]$, $a \sim x$. By transitivity, $a \sim x$ and $x \sim z$ implies $a \sim z$. Again by transitivity, $a \sim z$ and $z \sim y$ implies $a \sim y$. Hence, $a \in [y]$. It follows that

$[x] \subseteq [y]$. By reversing the roles of $x$ and $y$, we obtain $[y] \subseteq [x]$. Therefore, $[x] = [y]$ if $[x] \cap [y] \neq \varnothing$, and the result follows. $\qquad\square$

An important example of an equivalence relation is found in linear algebra (cf. Axler 2015, Section 3.E). Let $V$ be a vector space over the field of scalars $\mathbf{F}$, and $U$ a subspace of $V$. We define a binary relation $\sim$ on $V$ by

$$u \sim v \qquad \text{if} \qquad u - v \in U.$$

If $u \sim v$, then vectors $u$ and $v$ are said to be *congruent modulo U*. It is readily verified that $\sim$ is an equivalence relation on $V$. By Theorem 1.16, equivalence classes of $\sim$ form a partition of the vector space $V$. These classes are called *affine subspaces* of $V$. It can be verified that $[v] = U + v$, that is, an affine subspace $[v]$ is the translation of the subspace $U$ by $v$. The quotient set $V/\sim$ is usually denoted by $V/U$ and called the *quotient space of V modulo U*. The set $V/U$ is a vector space over the field $\mathbf{F}$ under the operations of addition and multiplication by a scalar defined by

$$[u] + [v] = [u] + [v] \qquad \text{and} \qquad k[u] = [ku],$$

for $u, v \in V$ and $k \in \mathbf{F}$. (The reader is invited to verify that these operations are well defined.)

## 1.5 Zorn's Lemma

It is customary to use one symbol for various order relations, a symbol that resembles the familiar inequality sign $\leq$. In what follows, we use the symbol $\preceq$ for these relations.

A *partial order* on a set $X$ is a reflexive, antisymmetric, and transitive binary relation $\preceq$ on $X$, that is, one that satisfies the following conditions for all $x, y, z \in X$:

| | |
|---|---|
| $x \preceq x$ | *reflexivity*, |
| if $x \preceq y$ and $y \preceq x$, then $x = y$ | *antisymmetry*, |
| if $x \preceq y$ and $y \preceq z$, then $x \preceq z$ | *transitivity*. |

If for every $x$ and $y$ in $X$ either $x \preceq y$ or $y \preceq x$, then the partial order $\preceq$ is called a *linear order* (also known as a *simple* or *total order*). A prototypical example of a linear order is the usual order $\leq$ on the set $\mathbf{N}$ of natural numbers (and on every nonempty set of real numbers).

*Example 1.4.* Let $X$ be a set. The set inclusion relation $\subseteq$ is a partial (and not always linear; cf. Exercise 1.15) order on the power set $\mathcal{P}(X)$ (the set of all subsets of the set $X$). It is a linear order if and only if $X = \varnothing$ or $X$ is a singleton.

*Example 1.5.* Let us define a binary relation $\preceq$ on the set $\mathbf{C}$ of complex numbers by

$$u \preceq v \quad \text{if and only if} \quad \mathrm{Re}(u) \leq \mathrm{Re}(v) \quad \text{and} \quad \mathrm{Im}(u) \leq \mathrm{Im}(v).$$

It is not difficult to verify (cf. Exercise 1.16) that this relation is indeed a partial (but not linear) order on $\mathbf{C}$. For instance, complex numbers 1 and $i$ are "incomparable" with respect to the relation $\preceq$ (that is, $1 \not\preceq i$ and $i \not\preceq 1$).

*Example 1.6.* (Cf. Example 1.5). Let $X = \mathbf{R}^2$ (the 2-dimensional real vector space) and define $\preceq$ on $X$ by

$$(x_1, y_1) \preceq (x_2, y_2) \quad \text{if either} \quad x_1 < x_2, \text{ or } x_1 = x_2 \text{ and } y_1 \leq y_2.$$

It can be seen that $\preceq$ is a linear order on $X$. Because of its resemblance to the way words are arranged in a dictionary, it is called the *lexicographical order* on $\mathbf{R}^2$.

*Example 1.7.* Let $X$ and $Y$ be sets and $F$ the set of all functions whose domain is a subset of $X$ and range is a subset of $Y$. For $f \in F$ we denote by $\mathcal{D}(f)$ the domain of the function $f$. We define a relation $\preceq$ on $F$ by

$$f \preceq g \quad \text{if} \quad \mathcal{D}(f) \subseteq \mathcal{D}(g) \text{ and } f(x) = g(x) \text{ for all } x \in \mathcal{D}(f).$$

In words: $f \preceq g$ means that $f$ is a restriction of $g$, or equivalently, that $g$ is an extension of $f$. Because functions in $F$ are subsets of the Cartesian product $X \times Y$, we observe that $\preceq$ on $F$ is an instance of the set inclusion relation.

A *partially ordered set* (a.k.a. *poset*) is a nonempty set together with a partial order on it. More formally, a partially ordered set is an ordered pair $(X, \preceq)$, where $X$ is a set and $\preceq$ is a partial order on $X$. The pair $(X, \preceq)$ is an example of an *algebraic structure*, a concept that is common in mathematics. It is customary to call the set $X$ itself a "partially ordered set." If the relation $\preceq$ is a linear order on $X$, then $X$ is often called a *chain*. Every nonempty subset $Y$ of a partially ordered set $(X, \preceq)$ is itself a partially ordered set with respect to the restriction of the relation $\preceq$ to $Y$.

Let $X$ be a partially ordered set. If there is an element $a \in X$ such that $a \preceq x$ for every $x$ in $X$, then $a$ is called a *least* element of $X$. By the antisymmetry property of order, if $X$ has a least element, then it is unique. Similarly, an element $a \in X$ is called a *greatest* element of $X$ if $x \preceq a$ for every $x$ in $X$; it also is unique if it exists.

*Example 1.8.* (1) The set of natural numbers $\mathbf{N} = \{1, 2, \ldots\}$ ordered by the usual $\leq$ relation has a least element but does not have a greatest element.

(2) The power set $\mathcal{P}(X)$ ordered by the inclusion relation has least element $\varnothing$ and greatest element $X$. These elements are equal only if $X = \varnothing$.

(3) Let $\{x_1, x_2, \ldots, x_n\}$ be a collection of distinct elements of some set $X$. The sets

$$\{x_1\}, \{x_1, x_2\}, \ldots, \{x_1, x_2, \ldots, x_n\}$$

form a chain

$$\{x_1\} \subseteq \{x_1, x_2\} \subseteq \cdots \subseteq \{x_1, x_2, \ldots, x_n\}$$

in the power set $\mathcal{P}(X)$.

If $Y$ is a subset of a partially ordered set $X$ and $a$ is an element of $X$ such that $y \preceq a$ for all $y \in Y$, then $a$ is said to be an *upper bound* of $Y$. A *lower bound* of a subset of $X$ is defined similarly.

An element $a$ of a partially ordered set $X$ is said to be a *minimal* element of $X$ if for every $x \in X$, $x \preceq a$ implies $x = a$. Similarly, an element $a$ is called a *maximal* element of $X$ if $a \preceq x$ implies $x = a$ for every $x \in X$.

*Example 1.9.* Let $X$ be a set containing more that one element. If the set $\mathcal{C} = \mathcal{P}(X) \setminus \varnothing$ of all nonempty subsets of $X$ is ordered by the inclusion relation, then each singleton is a minimal element of $\mathcal{C}$. However, $\mathcal{C}$ has no least element. The empty set is a lower bound of $\mathcal{C}$.

The following theorem is a fundamental result of set theory. The proof uses the axiom of choice and is omitted (cf. Halmos 1960, Chapter 16).

**Theorem 1.17. (Zorn's lemma.)** *If $X$ is a partially ordered set such that every chain in $X$ has an upper bound, then $X$ contains a maximal element.*

To exemplify applications of Zorn's lemma, we show that every nontrivial vector space has a Hamel basis.

First, we recall some basic concepts of linear algebra. Let $V$ be a vector space over a field $\mathbf{F}$. For a finite sequence of vectors $v_1, \ldots, v_n$ in $V$ and a finite sequence of scalars $\lambda_1, \ldots, \lambda_n$ in $\mathbf{F}$, an expression of the form

$$\lambda_1 v_1 + \cdots + \lambda_n v_n$$

is said to be a *linear combination* of vectors $v_1, \ldots, v_n$. Let $M$ be a subset of $V$. The *span* of $M$, denoted by span$(M)$, is the set of all linear combinations of vectors in $M$. A set $M \subseteq L$ is said to be *linearly independent* if for every sequence $v_1, \ldots, v_n$ of distinct vectors in $M$, the relation

$$\lambda_1 x_1 + \cdots + \lambda_n x_n = 0$$

implies $\lambda_k = 0$ for all $k = 1, \ldots, n$. A subset $B$ of vectors in $V$ is called a *Hamel basis* for $V$ if $B$ is linearly independent and spans $V$.

**Theorem 1.18.** *Every vector space $V$ has a Hamel basis.*

*Proof.* Let $\mathcal{M}$ be the set of all linearly independent subsets of $V$. This set is not empty, because it contains the empty set, which is linearly independent. The set inclusion relation defines a partial order on $\mathcal{M}$. Every chain $\mathcal{C}$ in $\mathcal{M}$ has an upper bound, namely the union $\bigcup \mathcal{C}$. By Zorn's lemma, $\mathcal{M}$ has a maximal element $B$. Suppose that $\operatorname{span}(B) \neq V$ and let $v$ be a vector in $V$ such that $v \notin \operatorname{span}(B)$. Let us show that $B \cup \{v\}$ is a linearly independent set. In the relation

$$\lambda_0 v + \lambda_1 v_1 + \cdots + \lambda_n v_n = 0,$$

where $v_k \in B$, $k = 1, \ldots, n$, we must have $\lambda_0 = 0$, because $v \notin \operatorname{span}(B)$. Inasmuch as $v_k \in B$, $k = 1, \ldots, n$, and $B$ is a linearly independent set, we have $\lambda_k = 0$, $k = 1, \ldots, n$. It follows that $B$ is a proper subset of the linearly independent set $B \cup \{v\}$, which contradicts the maximality of $B$.                    $\square$

## Notes

In Section 1.1 we present only properties of convex functions that are used in this chapter. More properties are found in Exercise 1.3. The reader can find comprehensive coverage of this important subject in many textbooks on real analysis (see, for instance, Wade 2000 and Royden and Fitzpatrick 2010).

If $f$ is a convex function on $\mathbf{R}$, $\varphi$ an integrable function over $[0, 1]$, and $f \circ \varphi$ also integrable over $[0, 1]$, then the following inequality holds:

$$f\left( \int_0^1 \varphi(x)\, dx \right) \leq \int_0^1 (f \circ \varphi)(x)\, dx.$$

This inequality is known as Jensen's inequality. The inequality in (1.3) is known as the finite form of Jensen's inequality.

Zorn's lemma, also known as the Kuratowski–Zorn lemma, was proved by Kazimierz Kuratowski in 1922 and independently by Max August Zorn in 1935. This lemma occurs in several fundamental theorems in mathematics. In addition to Theorem 1.18, we will use Zorn's lemma to establish the Hahn–Banach theorems in Chapter 5.

## Exercises

In Exercises 1.1–1.4, $I$ is an interval in $\mathbf{R}$.

**1.1.** For $x_1, \ldots, x_n \in I$, the number

$$x = \lambda_1 x_1 + \cdots + \lambda_n x_n$$

is said to be a *convex combination* of the numbers $x_1, \ldots, x_n$ if $0 \le \lambda_k \le 1$ for $1 \le k \le n$ and $\lambda_1 + \cdots + \lambda_n = 1$. Show that

$$\min\{x_1, \ldots, x_n\} \le x \le \max\{x_1, \ldots, x_n\}.$$

Accordingly, $x \in I$.

**1.2.** Prove that a function $f : I \to \mathbf{R}$ is convex if and only if for every three consecutive points $P$, $R$, $Q$ on the graph of the function, the point $R$ lies below or on the chord $PQ$ (cf. Fig. 1.1).

**1.3.** (a) Let $f$ be a real-valued function on $I$. Prove that $f$ is convex if and only if for all numbers $a$, $b$, and $x$ in $I$ such that $a < x < b$, the following inequalities hold:

$$\frac{f(x) - f(a)}{x - a} \le \frac{f(b) - f(a)}{b - a} \le \frac{f(b) - f(x)}{b - x}$$

(cf. Fig. 1.3).
(b) Show that

 (1) a convex function on $I$ has left-hand and right-hand derivatives, $f'(x^-)$ and $f'(x^+)$, at each point $x \in I$;
 (2) these derivatives are increasing functions on $I$; and
 (3) $f'(x^-) \le f'(x^+)$ on $I$.

(c) Prove Theorem 1.2.

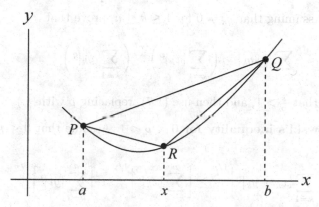

**Fig. 1.3** slope($PR$) $\le$ slope($PQ$) $\le$ slope($RQ$).

**1.4.** A function on $I$ of the form $f(x) = px + q$, $p, q \in \mathbf{R}$, is said to be *affine*. Show that

(a) $f$ is affine on $\mathbf{R}$ if and only if

$$f(\lambda x + (1 - \lambda)y) = \lambda f(x) + (1 - \lambda)f(y)$$

for all real numbers $x$ $y$, and $\lambda$.

(b) $f$ is affine on $I$ if and only if $f$ and $-f$ are convex on $I$.

(c) If $f$ is convex on $I = [a, b]$ and there exists $\lambda \in (0, 1)$ for which

$$f(\lambda a + (1 - \lambda)b) = \lambda f(a) + (1 - \lambda)f(b),$$

then $f$ is affine on $[a, b]$.

**1.5.** Show that for all $u, v, w \in \mathbf{C}$,

(a) $|u - w| \le |u - v| + |v - w|$.

(b) $|u - v| \ge ||u| - |v||$.

(c) $|u + v|^2 + |u - v|^2 = 2(|u|^2 + |v|^2)$.

**1.6.** Show that

$$\sum_{k=1}^{n} |x_k| \le \sqrt{n} \cdot \sqrt{\sum_{k=1}^{n} |x_k|^2}.$$

**1.7.** Prove inequalities (1.10) and (1.12).

**1.8. Hölder's inequality for $0 < p < 1$.** Let $p$ and $q$ be real numbers such that

$$0 < p < 1 \quad \text{and} \quad \frac{1}{p} + \frac{1}{q} = 1$$

(so $q < 0$). Assuming that $y_k \ne 0$ for $1 \le k \le n$, prove that

$$\sum_{k=1}^{n} |x_k y_k| \ge \left(\sum_{k=1}^{n} |x_k|^p\right)^{1/p} \left(\sum_{k=1}^{n} |y_k|^q\right)^{1/q}.$$

(Hint: Note that $\frac{1}{p} > 1$, and then use (1.7), replacing $p$ with $\frac{1}{p}$.)

**1.9. Minkowski's inequality for $0 < p < 1$.** Assume that $0 < p < 1$ and show that

$$\left[\sum_{k=1}^{n} |x_k + y_k|^p\right]^{1/p} \ge \left[\sum_{k=1}^{n} |x_k|^p\right]^{1/p} + \left[\sum_{k=1}^{n} |y_k|^p\right]^{1/p}.$$

**1.10.** Let $x_1, \ldots, x_n$ be positive real numbers. The classical Pythagorean means are the *arithmetic mean* ($A$), the *geometric mean* ($G$), and the *harmonic mean* ($H$). They are defined by

$$A = A(x_1, \ldots, x_n) = \frac{1}{n}(x_1 + \cdots + x_n),$$
$$G = G(x_1, \ldots, x_n) = \sqrt[n]{x_1 x_2 \cdots x_n},$$
$$H = H(x_1, \ldots, x_n) = \frac{n}{\frac{1}{x_1} + \cdots + \frac{1}{x_n}},$$

respectively. Prove that

$$\min\{x_1, \ldots, x_n\} \le H \le G \le A \le \max\{x_1, \ldots, x_n\}.$$

**1.11.** Let $a$, $b$, $x$, and $y$ be positive real numbers. Show that

$$x \ln \frac{x}{a} + y \ln \frac{y}{b} \ge (x+y) \ln \frac{x+y}{a+b}.$$

Hint: The function $f(t) = t \ln t$ is convex over $(0, \infty)$.

**1.12.** Prove Theorems 1.15 and 1.17.

**1.13.** Prove that the relation $\sim$ on the set $\mathbf{N} \times \mathbf{N}$ defined by

$$(m, n) \sim (p, q) \qquad \text{if} \qquad m + q = p + n,$$

for all $m, n, p, q \in \mathbf{N}$, is an equivalence relation. Describe the equivalence classes of this relation.

**1.14.** Prove that the relation $\sim$ in Example 1.3 is an equivalence relation on the set $\mathbf{Z} \times (\mathbf{Z} \setminus \{0\})$.

**1.15.** Let $X$ be a set with more than one element. Give examples of subsets $A$ and $B$ of $X$ such that $A \not\subseteq B$ and $B \not\subseteq A$.

**1.16.** Show that the relation $\preceq$ in Example 1.5 is a partial order on $\mathbf{C}$.

**1.17.** Consider the set $X$ of all closed disks in the plane that are subsets of the square $S = [0, 1] \times [0, 1]$.

(a) Describe the maximal elements of the poset $(X, \subseteq)$.
(b) Show that $(X, \subseteq)$ does not have a greatest element.
(c) Describe the minimal elements of $(X, \subseteq)$.
(d) Show that $(X, \subseteq)$ does not have a least element.

# Chapter 2
# Metric Spaces

The purpose of this chapter is to present a summary of some basic properties of metric and topological spaces that play an important role in the rest of the book.

## 2.1 Metrics and Pseudometrics

**Definition 2.1.** A *metric space* is a pair $(X, d)$, where $X$ is a nonempty set and $d$ is a function $d : X \times X \to \mathbf{R}$ satisfying the following conditions:

>   **(M1)** $d(x, y) = 0$ if and only if $x = y$,
>   **(M2)** $d(x, y) = d(y, x)$,            *symmetry*
>   **(M3)** $d(x, y) + d(y, z) \geq d(x, z)$,     *triangle inequality*

for all $x, y, z \in X$. Elements of the set $X$ are called *points*. The number $d(x, y)$ is said to be the *distance* between points $x$ and $y$. The function $d$ is called a *metric* or a *distance function*.

Note that conditions **M1–M3** imply that the distance function $d$ is necessarily nonnegative. Indeed,

$$0 = d(x, x) \leq d(x, y) + d(y, x) = 2\, d(x, y),$$

for all $x, y \in X$.

For any three points $x$, $y$, $z$ in a metric space $(X, d)$, the triangle inequality **(M3)** claims that each of the three distances $d(x, y)$, $d(y, z)$, and $d(z, x)$ is not greater than the sum of the other two distances:

$$d(x, y) \leq d(x, z) + d(z, y),$$
$$d(y, z) \leq d(y, x) + d(x, z),$$
$$d(z, x) \leq d(z, y) + d(y, x).$$

© Springer International Publishing AG, part of Springer Nature 2018
S. Ovchinnikov, *Functional Analysis*, Universitext,
https://doi.org/10.1007/978-3-319-91512-8_2

This system of three inequalities is equivalent to the following chain of inequalities:

$$|d(x,z) - d(y,z)| \leq d(x,y) \leq d(x,z) + d(z,y) \qquad (2.1)$$

(cf. Exercise 2.2).

By induction, the triangle inequality is generalized to $n > 3$ points as

$$d(x_1, x_n) \leq d(x_1, x_2) + \cdots + d(x_{n-1}, x_n) \qquad (2.2)$$

(cf. Exercise 2.4).

We will often refer to inequalities (2.1) and (2.2) as "triangle inequalities." The name is motivated by theorems in Euclidean geometry (cf. Fig. 2.1).

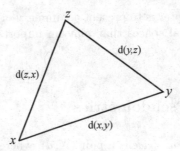

**Fig. 2.1** Triangle in the Euclidean plane.

*Example 2.1.* The set $\mathbf{R}$ of all real numbers endowed with the distance function

$$d(x,y) = |x - y|,$$

where $|x|$ is the absolute value of $x$, is a metric space. Similarly, the set of all complex numbers $\mathbf{C}$ is a metric space with the distance function

$$d(z,w) = |z - w|,$$

where $|z|$ is the modulus (absolute value) of $z$ in $\mathbf{C}$ (cf. (1.5)).

*Example 2.2.* Let $X$ be a nonempty set. It is easily seen that the function

$$d(x,y) = \begin{cases} 0, & \text{if } x = y, \\ 1, & \text{if } x \neq y, \end{cases}$$

is a metric, called the *discrete* metric (also known as the *trivial* metric) on $X$. The space $(X, d)$ is called the *discrete metric space*.

*Example 2.3.* Let $X$ be a vector space over the field $\mathbf{F}$ (which is either $\mathbf{R}$ or $\mathbf{C}$). A *norm* on $X$ is a real-valued function on $X$ whose value at $x \in X$ is denoted by $\|x\|$, with the following properties:

(a) $\|x\| = 0$ if and only if $x = 0$,
(b) $\|\alpha x\| = |\alpha| \|x\|$,
(c) $\|x + y\| \le \|x\| + \|y\|$,

for all $x, y \in X$ and $\alpha \in \mathbf{F}$. Note that these conditions imply that the norm is a nonnegative function on $X$. It can be readily verified that $d(x, y) = \|x - y\|$ is a metric on $X$.

Let $(X, d)$ be a metric space. It is clear that for a nonempty subset $Y$ of $X$ the restriction of $d$ to the set $Y \times Y$ is a metric on $Y$. Usually this metric is denoted by the same symbol $d$. The metric space $(Y, d)$ is called a *subspace* of the space $(X, d)$.

*Example 2.4.* Every nonempty set of real numbers is a metric space with the distance function given by $|x - y|$ (cf. Example 2.1). In particular, every interval in $\mathbf{R}$ is a metric space, as is the set $\mathbf{Q}$ of all rational numbers.

It is customary to omit reference to the metric $d$ in the notation $(X, d)$ and write "a metric space $X$" instead of "a metric space $(X, d)$." However, we use the latter notation if different metrics on the same set are considered.

A concept of distance that is more general than that of a metric is especially useful in functional analysis.

**Definition 2.2.** A *pseudometric* on a nonempty set $X$ is a real-valued function $d$ satisfying the following conditions:

    **(PM1)** $d(x, x) = 0$,
    **(PM2)** $d(x, y) = d(y, x)$,         *symmetry*
    **(PM3)** $d(x, y) + d(y, z) \ge d(x, z)$,    *triangle inequality*

for all $x, y, z \in X$ (cf. Definition 2.1). The pair $(X, d)$ is said to be a *pseudometric space*.

A metric space is clearly also a pseudometric space such that the distance between two distinct points is always a positive number.

*Example 2.5.* Consider the following two functions on the set $\mathbf{C}$:

$$d_1(z, w) = |\mathrm{Re}(z) - \mathrm{Re}(w)| \quad \text{and} \quad d_2(z, w) = |\mathrm{Im}(z) - \mathrm{Im}(w)|.$$

Both functions are pseudometrics on $\mathbf{C}$. Note, for instance, that $d_2(x, y) = 0$ for all real numbers $x$ and $y$.

*Example 2.6.* Let $X$ be a vector space over the field $\mathbf{F}$. A function $p : X \to \mathbf{R}$ is called a *seminorm on $X$* if it satisfies the following conditions:

(a) $p(\alpha x) = |\alpha| p(x)$,

(b) $p(x + y) \le p(x) + p(y)$,

for all $\alpha \in \mathbf{F}$ and $x, y \in X$. It can be shown (cf. Exercise 2.7) that the function

$$d(x, y) = p(x - y)$$

is a pseudometric on $X$.

## 2.2 Open and Closed Sets

**Definition 2.3.** Let $X$ be a metric space, $x \in X$, and $r > 0$. The set

$$B(x, r) = \{y \in X : d(y, x) < r\}$$

is called the *open ball of radius $r$ centered at $x$*. Similarly, the *closed ball of radius $r$ centered at $x$* is the set

$$\overline{B}(x, r) = \{y \in X : d(y, x) \le r\}.$$

The definitions of open and closed balls have their roots in Euclidean geometry. However, in abstract metric spaces, these concepts may have some counterintuitive properties (cf. Exercises 2.11 and 2.13).

The open ball of radius $\varepsilon > 0$ centered at $x \in X$ is called the *$\varepsilon$-neighborhood* of $x$. A *neighborhood* of $x \in X$ is a subset of the space $X$ containing the $\varepsilon$-neighborhood of $x$ for some $\varepsilon > 0$.

Let $E$ be a subset of a metric space $X$. A point $x \in E$ is said to be an *interior point of $E$* if $E$ contains an open ball centered at $x$. The *interior of the set $E$* is the set of all its interior points. This set is denoted by int $E$.

*Example 2.7.* In these examples, all sets under consideration are subsets of the metric space $\mathbf{R}$.

(a) The interior of an open interval $(a, b)$ is the interval itself.

(b) The interior of the closed interval $[0, 1]$ is the open interval $(0, 1)$.

(c) The interior of the set of rational numbers $\mathbf{Q}$ is empty (cf. Exercise 2.16).

**Definition 2.4.** A subset $U$ of a metric space $X$ is said to be *open* if it contains an open ball centered at each of its points.

In other words, a subset $U$ of $X$ is an open set if it coincides with its interior.

Every metric space $X$ has at least two distinct open subsets, namely the empty set and the set $X$ itself. If the metric space $X$ consists of a single point, then $\varnothing$ and $X$ are the only open subsets of $X$ (cf. Exercise 2.17).

Two fundamental properties of open sets in a metric space are found in the next theorem.

**Theorem 2.1.** *Let $X$ be a metric space and $\mathcal{T}$ the collection of all open subsets of $X$. Then:*

(a) *The union of every family $\{U_i\}_{i \in J}$ of open sets is an open set:*

$$\bigcup_{i \in J} U_i \in \mathcal{T}.$$

(b) *The intersection of every finite family $\{U_1, \ldots, U_n\}$ of open sets is an open set:*

$$\bigcap_{i=1}^{n} U_i \in \mathcal{T}.$$

*Proof.* (a) If a point $x$ belongs to the union $\bigcup_{i \in J} U_i$, then there is $i \in J$ such that $x \in U_i$. Because $U_i$ is an open set, there is an open ball $B(x, r)$ that is a subset of $U_i$. Clearly, $B(x, r) \subseteq \bigcup_{i \in J} U_i$. Hence, $\bigcup_{i \in J} U_i$ is an open set.

(b) By induction, it suffices to show that the intersection $U_1 \cap U_2$ of two open sets is open. If this intersection is empty, then we are done, because the empty set is open. Otherwise, let $x \in U_1 \cap U_2$. Inasmuch as the sets $U_1$ and $U_2$ are open, there are balls $B(x, r_1) \subseteq U_1$ and $B(x, r_2) \subseteq U_2$. Then the ball $B(x, r)$ with $r = \min\{r_1, r_2\}$ belongs to the intersection $U_1 \cap U_2$, which is the desired result. $\qquad \square$

Let $E$ be a subset of a metric space $X$. A point $x \in X$ is said to be a *limit point of $E$* (or an *accumulation point of $E$*) if every neighborhood of $x$ contains a point in $E$ distinct from $x$. The set consisting of the points of $E$ and the limit points of $E$ is called the *closure* of $M$ and is denoted by $\overline{E}$.

**Definition 2.5.** A subset $F$ of a metric space $X$ is said to be *closed* if it contains all its limit points.

Hence, a subset $F$ of $X$ is closed if it coincides with its closure, $F = \overline{F}$.

*Example 2.8.* (a) The open unit ball in the complex plane,

$$B(0, 1) = \{z \in \mathbf{C} : |z| < 1\},$$

is an open set.

(b) The closed unit ball in the complex plane,

$$\overline{B}(0, 1) = \{z \in \mathbf{C} : |z| \leq 1\},$$

is a closed set, which is the closure of $B(0, 1)$,

$$\overline{B(0, 1)} = \overline{B}(0, 1).$$

(c) Let $X$ be a discrete metric space of cardinality greater than 1 and $x \in X$. Then

$$B(x,1) = \{x\}, \qquad \overline{B(x,1)} = \{x\}, \qquad \text{and} \qquad \overline{B}(x,1) = X \neq \{x\},$$

so the closure of an open ball $B(x,r)$ in a metric space can be a proper subset of the closed ball $\overline{B}(x,r)$ (cf Exercise 2.18).

The following theorem establishes a close relationship between the concepts of open and closed sets.

**Theorem 2.2.** *A subset $E$ of a metric space $X$ is closed if and only if its complement $E^c$ is open.*

*Proof.* It is easy to see that the result holds if $E = \varnothing$ or $E = X$. Thus we assume that $E$ is a nonempty subset of $X$ that is different from $X$.

(Necessity.) Let $E$ be a closed set and $x$ an element of $E^c = X \setminus E$. Because $x \notin E$, it is not a limit point of $E$. Hence, there is a neighborhood of $x$ that is a subset of $E^c$. It follows that $E^c$ is an open set.

(Sufficiency.) Suppose that the complement $E^c$ of the set $E$ is open and let $x$ be a limit point of $E$. Because every neighborhood of $x$ contains a point of $E$ different from $x$, the point $x$ does not belong to the open set $E^c$. Therefore, $x \in E$, that is, $E$ is a closed set.                                                    $\square$

By applying Theorem 2.2 to properties (a) and (b) of open sets established in Theorem 2.1, we obtain fundamental properties of closed sets.

**Theorem 2.3.** *In a metric space,*

(a) *the union of a finite family of closed sets is a closed set;*
(b) *the intersection of every family of closed sets is a closed set.*

The proof of the next theorem is left as an exercise (cf. Exercise 2.19).

**Theorem 2.4.** *The interior of a subset $E$ of a metric space is the maximum open set contained in $E$. The closure of $E$ is the minimum closed set containing $E$.*

A subset $E$ of a metric space $X$ is said to be *dense in $X$* if its closure is the entire space $X$:
$$\overline{E} = X.$$

If $E$ is dense in $X$, then every ball in $X$ contains a point from $E$.

A metric space is called *separable* if it has a countable dense subset.

*Example 2.9.* (a) The metric space of real numbers $\mathbf{R}$ is separable, because the set of rational numbers $\mathbf{Q}$ is countable and dense in $\mathbf{R}$.

(b) The space $\mathbf{C}$ of complex numbers is also separable (cf. Exercise 2.23).

(c) A discrete metric space is separable if and only if it is countable (cf. Exercise 2.24).

## 2.3 Convergence and Completeness

**Definition 2.6.** A point $x$ in a metric space $X$ is said to be a *limit* of a sequence of points $(x_n)$ in $X$ if for every $\varepsilon > 0$, there exists $N \in \mathbf{N}$ such that

$$d(x_n, x) < \varepsilon, \qquad \text{for all } n > N.$$

If $x$ is a limit of the sequence $(x_n)$, we say that $(x_n)$ *converges* to $x$ and write

$$x_n \to x.$$

If a sequence has a limit, it is called *convergent*. Otherwise, it is called *divergent*.

It is not difficult to see that

$$x_n \to x \qquad \text{if and only if} \qquad d(x_n, x) \to 0.$$

*Example 2.10.* Every point $x$ in a metric space $X$ is a limit of the *constant sequence*: $x_n = x$, for all $n \in \mathbf{N}$.

*Example 2.11.* Let $X = [0, 1]$ be a subspace of $\mathbf{R}$. The sequence $(1/n)$ converges to $0$ in $X$. On the other hand, the same sequence diverges in the subspace $(0, 1)$ of $\mathbf{R}$ (cf. Exercise 2.35).

**Theorem 2.5.** *A sequence of points in a metric space has at most one limit.*

*Proof.* Suppose $(x_n)$ is a convergent sequence such that $x_n \to x$ and $x_n \to y$. Then

$$0 \le d(x, y) \le d(x, x_n) + d(x_n, y) \to 0.$$

Hence, $d(x, y) = 0$, so $x = y$. □

This theorem justifies the notation

$$\lim_{n \to \infty} x_n = x \qquad \text{(or simply} \lim x_n = x)$$

for the limit of a sequence $(x_n)$ that converges to $x$.

**Definition 2.7.** A sequence of points $(x_n)$ in a metric space $X$ is said to be a *Cauchy sequence* (or *fundamental sequence*) if for every $\varepsilon > 0$, there exists $N \in \mathbf{N}$ such that

$$d(x_m, x_n) < \varepsilon, \qquad \text{for all } m, n > N.$$

**Theorem 2.6.** *Every convergent sequence in a metric space is a Cauchy sequence.*

*Proof.* Suppose $x_n \to x$ in a metric space $X$. Then for a given $\varepsilon > 0$, there exists $N \in \mathbf{N}$ such that

$$d(x_n, x) < \frac{\varepsilon}{2}, \qquad \text{for all } n > N.$$

By the triangle inequality, we have

$$d(x_m, x_n) \le d(x_m, x) + d(x, x_n) < \frac{\varepsilon}{2} + \frac{\varepsilon}{2} = \varepsilon,$$

for all $m, n > N$. Hence $(x_n)$ is a Cauchy sequence.                               $\square$

The converse of this theorem is not true in a general metric space. For instance, the sequence $(1/n)$ in the space $X = (0, 1)$ (cf. Exercise 2.35) is Cauchy but does not converge to a point in $X$.

A subset $E$ of a metric space is said to be *bounded* if there exists $M > 0$ such that $d(x, y) < M$ for all $x, y \in E$. A sequence $(x_n)$ of points in a metric space is *bounded* if the set $\{x_n : n \in \mathbf{N}\}$ is bounded.

**Theorem 2.7.** *Every Cauchy sequence in a metric space is bounded. In particular, every convergent sequence in a metric space is bounded.*

*Proof.* Let $(x_n)$ be a Cauchy sequence in a metric space $X$. Then for $\varepsilon = 1$, there exists $N \in \mathbf{N}$ such that

$$d(x_m, x_n) < 1, \qquad \text{for all } m, n \ge N.$$

Let $a = \max_{1 \le k < N}\{d(x_N, x_k)\}$. Then $d(x_N, x_n) < \max\{1, a\}$, for all $n \in \mathbf{N}$. By the triangle inequality,

$$d(x_m, x_n) \le d(x_m, x_N) + d(x_N, x_n) < 2\max\{1, a\}, \qquad \text{for all } m, n \in \mathbf{N}.$$

Hence the sequence $(x_n)$ is bounded.

By Theorem 2.6, every convergent sequence in $X$ is bounded.                          $\square$

Metric spaces in which Cauchy sequences converge are important in analysis and deserve a name.

**Definition 2.8.** A metric space $X$ is said to be *complete* if every Cauchy sequence in $X$ converges (that is, has a limit in $X$). Otherwise, we say that $X$ is *incomplete*.

*Example 2.12.* The metric spaces $\mathbf{R}$ and $\mathbf{C}$ are complete. A closed interval in $\mathbf{R}$ and a closed disk in $\mathbf{C}$ are also complete metric spaces. An open interval in $\mathbf{R}$ and an open disk in $\mathbf{C}$ are examples of incomplete metric spaces.

The following theorem gives a criterion for a subspace of a complete metric space to be itself a complete space (cf. the previous example).

**Theorem 2.8.** *A subspace $Y$ of a complete space $X$ is complete if and only if $Y$ is a closed subset of $X$.*

*Proof.* (Necessity.) Suppose that $Y$ is complete and let $y \in X$ be a limit point of $Y$. Then for every $n \in \mathbf{N}$, there exists $y_n \in Y$ such that $d(y_n, y) < \frac{1}{n}$. Because $y_n \to y$, the sequence $(y_n)$ is Cauchy in $Y$. Inasmuch as $Y$ is complete, we have $y \in Y$. Hence $Y$ is a closed subset of $X$.

(Sufficiency.) Let $Y$ be a closed subset of $X$ (so $\overline{Y} = Y$) and let $(y_n)$ be a Cauchy sequence of points in $Y$. Because this sequence is also a sequence of points in $X$, and $X$ is complete, $(y_n)$ converges to a point $x \in X$. If $x \in Y$, we are done. Otherwise, $x$ is a limit point of $Y$, so $x \in \overline{Y} = Y$. □

An important example of an incomplete space is the metric space $\mathbf{Q}$ with the distance function $d(x, y) = |x - y|$ for $x, y \in \mathbf{Q}$. Of course, this follows from the previous theorem, because $\mathbf{Q}$ is not a closed subset of the complete space $\mathbf{R}$. To show that $\mathbf{Q}$ is an incomplete metric space directly, it suffices to produce an example of a divergent Cauchy sequence in this space.

*Example 2.13.* Consider the sequence of rational numbers

$$s_n = 1 - \frac{1}{2!} - \frac{1}{3!} + \cdots + \frac{(-1)^{n+1}}{n!} = \sum_{k=1}^{n} \frac{(-1)^{k+1}}{k!}, \qquad n \in \mathbf{N}.$$

First, we show that $(s_n)$ is a Cauchy sequence in $\mathbf{Q}$. For $m > n$, we have

$$|s_m - s_n| = \left| \frac{(-1)^{n+2}}{(n+1)!} + \cdots + \frac{(-1)^{m+1}}{m!} \right| \leq \frac{1}{2^{n-1}} + \cdots + \frac{1}{2^{m-2}} < \frac{1}{2^n}.$$

(We use the inequality $n! > 2^{n-2}$ for all $n \in \mathbf{N}$, and the sum of a geometric series.) It follows that $(s_n)$ is Cauchy.

Suppose that $(s_n)$ converges to a rational number $r$, that is,

$$\sum_{k=1}^{\infty} \frac{(-1)^{k+1}}{k!} = r.$$

Because the series on the left-hand side satisfies the alternating series test conditions, we have the following estimate for the remainder (cf. Exercise 2.36):

$$0 < |r - s_n| < \frac{1}{(n+1)!}, \qquad \text{for } n \in \mathbf{N}.$$

Let $r = \dfrac{p}{q} = \dfrac{p(q-1)!}{q!}$ and $n = q$. From the above inequalities, we have

$$0 < \left| \frac{p(q-1)!}{q!} - \sum_{k=1}^{q} \frac{(-1)^{k+1}}{k!} \right| < \frac{1}{(q+1)!}.$$

By multiplying all sides of these inequalities by $q!$, we obtain a contradiction,

$$0 < \left| p(q-1)! - \sum_{k=1}^{q} \frac{(-1)^{k+1} q!}{k!} \right| < \frac{1}{q+1},$$

because the number in the middle is an integer. It follows that the Cauchy sequence $(s_n)$ diverges in $\mathbf{Q}$.

The *diameter* of a bounded set $A$ in a metric space $(X, d)$ is defined as

$$\mathrm{diam}(A) = \sup\{d(a, b) : a, b \in A\}.$$

A nested family $A_1 \supseteq A_2 \supseteq \cdots$ of nonempty subsets of $X$ is said to be *contracting* if

$$\mathrm{diam}(A_n) \to 0.$$

The following property of complete metric spaces is useful in applications.

**Theorem 2.9. (The Cantor intersection property)** *If a metric space $X$ is complete, then whenever $A_1 \supseteq A_2 \supseteq \cdots$ is a contracting sequence of nonempty closed subsets of $X$, there exists $x \in X$ for which*

$$\bigcap_{k=1}^{\infty} A_k = \{x\}.$$

*Proof.* Let $r_n = \mathrm{diam}(A_n)$, for $n \in \mathbf{N}$. The sequence $(r_n)$ is decreasing and converges to zero. For each $n \in \mathbf{N}$, choose $x_n \in A_n$. For a given $\varepsilon > 0$, there exists $N \in \mathbf{N}$ such that $r_n < \varepsilon$ for all $n > N$. Hence for all $m, n > N$, $d(x_m, x_n) < \varepsilon$, so $(x_n)$ is Cauchy. Therefore, there exists $x \in X$ such that $x_n \to x$. Inasmuch as the sets $A_n$ are closed, we have $x \in A_n$ for all $n \in \mathbf{N}$, so $x \in \bigcap_{k=1}^{\infty} A_k$. The claim follows because $r_n \to 0$. $\qquad\square$

*Example 2.14.* We show that the set of real numbers $\mathbf{R}$ is uncountable. The proof is by contradiction, so we suppose that there is a sequence of real numbers

$$x_1, x_2, \ldots, x_n, \ldots$$

such that every real number is a term of this sequence.

It is clear that there is a closed bounded interval $[a_1, b_1]$ of length

$$b_1 - a_1 = \frac{1}{3}$$

such that $x_1 \notin [a_1, b_1]$. We divide this interval into three closed subintervals of equal length. Again, it is clear that there is at least one of these subintervals, say $[a_2, b_2]$, that does not contain $x_2$ (and does not contain $x_1$ either). We construct recursively a nested family of closed intervals

$$\{[a_n, b_n] : n \in \mathbf{N}\}$$

such that for all $n \in \mathbf{N}$,

$$b_n - a_n = \frac{1}{3^n},$$

and $[a_n, b_n]$ does not contain the numbers $x_1, \ldots, x_n$. By Theorem 2.9, there is a real number $a$ that belongs to all intervals $[a_n, b_n]$ and hence is not a term of the sequence $(x_n)$.

## 2.4 Mappings

Let $X$ and $Y$ be sets. Recall that a *function* $f : X \to Y$ is a subset $f \subseteq X \times Y$ satisfying the following condition: for every $x \in X$ there is a unique $y \in Y$ such that $(x, y) \in f$. Then we write $y = f(x)$ and call $X$ the *domain* of the function $f$ and the set $Y$ its *codomain*.

In functional analysis, we often use different names instead of "function," such as "mapping," "transformation," and "operator," among other variations. Some notations are also abbreviated to make formulas more readable.

The following definition illustrates some of these conventions. It is clear that this definition generalizes the concept of continuity in elementary analysis.

**Definition 2.9.** Let $(X, d)$ and $(Y, d')$ be metric spaces. A mapping (that is, a function) $T : X \to Y$ is said to be *continuous at* $x_0 \in X$ if for every $\varepsilon > 0$ there is $\delta > 0$ such that

$$d(x, x_0) < \delta \qquad \text{implies} \qquad d'(Tx, Tx_0) < \varepsilon, \qquad \text{for all } x \in X.$$

The mapping $T$ is said to be *continuous* (on $X$) if it is continuous at every point of $X$.

Note that we write $Tx$ instead of the usual notation $T(x)$ for the value of a function at $x$.

The next two theorems give equivalent forms of this definition.

**Theorem 2.10.** *A mapping $T : X \to Y$ is continuous at $x_0 \in X$ if and only if $Tx_n \to Tx_0$ for every sequence $(x_n)$ in $X$ such that $x_n \to x_0$.*

*Proof.* (Necessity.) Let $T : X \to Y$ be a continuous mapping and $x_n \to x_0$ in the space $X$. For $\varepsilon > 0$ there exists $\delta > 0$ such that $d'(Tx, Tx_0) < \varepsilon$ whenever $d(x, x_0) < \delta$. Because $x_n \to x_0$, there exists $N \in \mathbf{N}$ such that $d(x_n, x_0) < \delta$ for $n > N$. Hence, $d'(Tx_n, Tx_0) < \varepsilon$ for $n > N$. It follows that $Tx_n \to Tx_0$ in the space $Y$.

(Sufficiency.) Suppose that $T$ is not continuous at $x_0$. Then there exists $\varepsilon > 0$ such that for every $n \in \mathbf{N}$, there exists $x_n \in X$ such that $d(x_n, x_0) < 1/n$

and $d'(Tx_n, Tx_0) \geq \varepsilon$. It is clear that $(Tx_n)$ does not converge to $Tx_0$ in $Y$, whereas $x_n \to x_0$ in $X$. This contradiction shows that $T$ is continuous at $x_0$. $\square$

**Theorem 2.11.** *A mapping* $T : X \to Y$ *is continuous on* $X$ *if and only if the inverse image of every open subset of* $Y$ *is an open subset of* $X$.

*Proof.* (Necessity.) Let $T : X \to Y$ be a continuous mapping and $U$ an open set in $Y$. For $x_0 \in T^{-1}(U)$ choose $\varepsilon > 0$ such that $B(Tx_0, \varepsilon) \subseteq U$. This is possible because $Tx_0 \in U$ and $U$ is an open set. Because $T$ is continuous, there exists $\delta > 0$ such that $d(x, x_0) < \delta$ implies $Tx \in B(Tx_0, \varepsilon)$. It follows that $B(x_0, \delta) \subseteq T^{-1}(U)$, so $T^{-1}(U)$ is an open subset of $X$.

(Sufficiency.) Suppose that the inverse image of every open set in $Y$ is an open set in $X$. Let $x_0$ be a point in $X$ and $\varepsilon > 0$. Let $V \subseteq X$ be the inverse image of the open ball $B(Tx_0, \varepsilon)$. Because $V$ is an open set and $x_0 \in V$, there is $\delta > 0$ such that $B(x_0, \delta) \subseteq V$, that is, $d(x, x_0) < \delta$ implies $d'(Tx, Tx_0) < \varepsilon$ for all $x \in X$. It follows that $T$ is continuous at $x_0$. $\square$

An example of a continuous function is the distance function on a metric space. This function is continuous in each of its arguments. Here is the precise formulation of this property.

**Theorem 2.12.** *Let* $(X, d)$ *be a metric space. If* $x_n \to x$ *and* $y_n \to y$ *in* $X$, *then*

$$d(x_n, y_n) \to d(x, y),$$

*or equivalently,*

$$d(\lim x_n, \lim y_n) = \lim d(x_n, y_n),$$

*provided that both limits on the left-hand side exist.*

*Proof.* The claim of the theorem follows from the "quadrilateral inequality":

$$|d(x_n, y_n) - d(x, y)| \leq d(x_n, x) + d(y_n, y)$$

(cf. Exercise 2.3). $\square$

Another example of a continuous mapping is an isometry.

**Definition 2.10.** Let $(X, d)$ and $(Y, d')$ be metric spaces. A mapping $T$ of $X$ into $Y$ is said to be an *isometry* if it preserves distances, that is,

$$d'(Tx, Ty) = d(x, y), \qquad \text{for all } x, y \in X.$$

The space $X$ is said to be *isometric* to the space $Y$ if there is a bijective isometry from $X$ onto $Y$. Then the spaces $X$ and $Y$ are called *isometric*.

It is not difficult to see that an isometry is indeed a continuous mapping. The claim of the next theorem is an important property of isometric spaces.

**Theorem 2.13.** *Let $(X, d)$ and $(X', d')$ be complete metric spaces. If $U$ and $U'$ are dense subsets of $X$ and $X'$, respectively, and $T$ is an isometry from $U$ onto $U'$, then there exists an isometry $T'$ from $X$ onto $X'$ such that $T = T'|_U$, that is, $T'$ is an extension of $T$ to $X$.*

*In other words, if $U$ and $U'$ are isometric dense subsets of complete metric spaces $X$ and $X'$, respectively, then the spaces $X$ and $X'$ are themselves isometric.*

*Proof.* Let $x$ be a point in $X$. Because $U$ is dense in $X$, there is a sequence $(u_n)$ of points of $U$ that converges to $x$. Suppose that $(v_n)$ is another sequence with the same properties. Since convergent sequence are Cauchy and $T$ preserves distances, the sequences $(Tu_n)$ and $(Tv_n)$ are also Cauchy. They are convergent because $X'$ is complete. By Theorem 2.12,

$$d'(\lim Tu_n, \lim Tv_n) = \lim d'(Tu_n, Tv_n) = \lim d(u_n, v_n)$$
$$= d(\lim u_n, \lim v_n) = d(x, x) = 0.$$

Hence $\lim Tu_n = \lim Tv_n$ for every two sequences in $U$ that converge to $x$. It follows that $T'x = \lim Tu_n$ is a well-defined function from $X$ into $X'$.

Now we show that $T'$ preserves distances. Let $(u_n)$ and $(v_n)$ be sequences of points in $U$ that converge to points $x$ and $y$ in $X$, respectively. We have, again by Theorem 2.12,

$$d'(T'x, T'y) = d'(\lim Tu_n, \lim Tv_n) = \lim d'(Tu_n, Tv_n)$$
$$= \lim d(u_n, v_n) = d(\lim u_n, \lim v_n) = d(x, y).$$

Hence, the mapping $T'$ is an isometry.

It remains to show that $T'$ is a surjective mapping. Let $x'$ be a point in $X'$. Because $U'$ is dense in $X'$, there is a sequence $(u'_n)$ of points of $U'$ that converges to $x'$. Because this sequence converges, it is Cauchy. Since $T^{-1}$ is an isometry, the sequence $T^{-1}u'_n$ is also a Cauchy sequence. Inasmuch as $X$ is a complete space, this sequence converges to some point $x$ in $X$, that is, $u_n \to x$, where $u_n = T^{-1}u'_n$. Hence,

$$T'x = \lim Tu_n = \lim u'_n = x',$$

and the result follows.                                                              □

## 2.5 Completion of a Metric Space

The completeness property of the field of real numbers $\mathbf{R}$ is crucial in the proofs of many theorems of real analysis. Complete metric spaces play an important role in many constructions of functional analysis. However, sometimes one has to deal with incomplete metric spaces. In this section, we describe a remarkable construction that makes it possible to create a unique complete space from an incomplete one. We begin with a definition.

**Definition 2.11.** A *completion* of a metric space $(X, d)$ is a metric space $(\widetilde{X}, \widetilde{d})$ with the following properties:

(a) $\widetilde{X}$ is a complete space, and
(b) $(X, d)$ is isometric to a dense subset of $(\widetilde{X}, \widetilde{d})$.

It is known from real analysis that the space of real numbers $\mathbf{R}$ is a completion of its subspace of rational numbers $\mathbf{Q}$.

**Theorem 2.14.** *For a metric space $(X, d)$ there exists its completion $(\widetilde{X}, \widetilde{d})$.*

*Proof.* First, we construct the space $(\widetilde{X}, \widetilde{d})$. Let $X'$ be the set of Cauchy sequences of elements of $X$. We define the function $d'((x_n), (y_n))$ on $X'$ by

$$d'((x_n), (y_n)) = \lim d(x_n, y_n).$$

This function is well defined. Indeed, we have

$$|d(x_m, y_m) - d(x_n, y_n)| \le d(x_m, x_n) + d(y_m, y_n), \qquad \text{for all } m, n \in \mathbf{N},$$

by the "quadrilateral inequality" (cf. Exercise 2.3). Because $(x_n)$ and $(y_n)$ are Cauchy sequences in $X$, the sequence of real numbers $d(x_n, y_n)$ is Cauchy and therefore convergent.

It is not difficult to verify that the function $d'$ is a pseudometric (cf. Exercise 2.4) on $X'$. Let $\widetilde{X} = X'/\sim$ be the quotient set of $X'$ with respect to the equivalence relation on $X'$ defined by

$$(x_n) \sim (y_n) \qquad \text{if and only if} \qquad d'((x_n), (y_n)) = 0.$$

According to Exercise 2.4, the set $\widetilde{X}$ equipped with the distance function

$$\widetilde{d}([(x_n)], [(y_n)]) = d'((x_n), (y_n))$$

is a metric space.

We proceed with the construction of an isometry $T : X \to \widetilde{X}$. With each $x \in X$ we associate the class $[(x)]$ of the constant sequence $(x)$ and claim that the mapping given by $Tx = [(x)]$ is an isometry from $X$ onto $T(X) \subseteq \widetilde{X}$. Indeed, for constant sequences $(x)$ and $(y)$, we have

$$\widetilde{d}([(x)], [(y)]) = d'((x), (y)) = d(x, y),$$

by the definition of the functions $\widetilde{d}$ and $d'$.

Now we show that $T(X)$ is dense in $\widetilde{X}$. For this we need to show that every open ball in $\widetilde{X}$ contains a point of $T(X)$. Let $[(x_n)] \in \widetilde{X}$. Because $(x_n)$ is a Cauchy sequence in $X$, for every $\varepsilon > 0$ there exists $N \in \mathbf{N}$ such that

$$d(x_n, x_N) < \frac{\varepsilon}{2}, \qquad \text{for all } n > N.$$

Consider the constant sequence $(x_N) = (x_N, x_N, \ldots)$ in $X$, so $[(x_N)] \in T(X)$. We have

$$\widetilde{d}([(x_n)], [(x_N)]) = d'((x_n), (x_N)) = \lim d(x_n, x_N) \le \frac{\varepsilon}{2} < \varepsilon.$$

Hence, the ball $B([(x_n)], \varepsilon)$ contains the point $[(x_N)]$ of $T(X)$.

It remains to show that $(\widetilde{X}, \widetilde{d})$ is a complete space. Let $(\widetilde{x}_1, \widetilde{x}_2, \ldots, \widetilde{x}_n, \ldots)$ be a Cauchy sequence in $\widetilde{X}$. Because $T(X)$ is dense in $\widetilde{X}$, it follows that for every $n \in \mathbf{N}$, there is a constant sequence $(z^{(n)}) = (z_n, z_n, \ldots)$ in $X$ such that

$$\widetilde{d}(\widetilde{x}_n, \widetilde{z}_n) < \frac{1}{n},$$

where $\widetilde{z}_n = [(z^{(n)})] \in T(X)$ is the class of the constant sequence $(z^{(n)})$. We have

$$\widetilde{d}(\widetilde{z}_m, \widetilde{z}_n) \le \widetilde{d}(\widetilde{z}_m, \widetilde{x}_m) + \widetilde{d}(\widetilde{x}_m, \widetilde{x}_n) + \widetilde{d}(\widetilde{x}_n, \widetilde{z}_n) < \frac{1}{m} + \widetilde{d}(\widetilde{x}_m, \widetilde{x}_n) + \frac{1}{n},$$

and the right-hand side can be made smaller than every given positive number for sufficiently large $m$ and $n$, because $(\widetilde{x}_n)$ is a Cauchy sequence. Hence $(\widetilde{z}_n)$ is Cauchy. Because $T z_n = [(z^{(n)})] = \widetilde{z}_n$ and $T$ is an isometry of $X$ onto $T(X)$, the sequence $(z_n) = (z_1, z_2, \ldots)$ is a Cauchy sequence in $X$. Let $\widetilde{x} = [(z_n)]$. We show that $\widetilde{x}_n \to \widetilde{x}$, which establishes completeness of $\widetilde{X}$.

We have

$$\widetilde{d}(\widetilde{x}_n, \widetilde{x}) \le \widetilde{d}(\widetilde{x}_n, \widetilde{z}_n) + \widetilde{d}(\widetilde{z}_n, \widetilde{x}) < \frac{1}{n} + \widetilde{d}(\widetilde{z}_n, \widetilde{x}) = \frac{1}{n} + d'((z^{(n)}), (z_n)).$$

Using the definition of $d'$, we obtain

$$\widetilde{d}(\widetilde{x}_n, \widetilde{x}) < \frac{1}{n} + \lim_{m \to \infty} d(z_n, z_m).$$

Because $(z_n)$ is a Cauchy sequence, the right-hand side of the above inequality can be made smaller than every given positive number for sufficiently large $n$, implying that $\widetilde{x}_n \to \widetilde{x}$. $\qquad\square$

The next theorem asserts that in some precise sense there is only one completion of a metric space.

**Theorem 2.15.** *If $(\widetilde{X}, \widetilde{d})$ and $(\widehat{X}, \widehat{d})$ are two completions of a metric space $(X, d)$, then the spaces $\widetilde{X}$ and $\widehat{X}$ are isometric.*

*Proof.* Let $T$ and $T'$ be isometries of $X$ onto dense sets $T(X)$ and $T'(X)$ in spaces $\widetilde{X}$ and $\widehat{X}$, respectively. Then $T' \circ T^{-1}$ is an isometry of $T(X)$ onto $T'(X)$. The claim of the theorem follows from Theorem 2.13.                    □

## 2.6 The Baire Category Theorem

The result that we establish in this section—the Baire category theorem—has powerful applications in functional analysis and elsewhere in mathematics. We begin with a simple lemma.

**Lemma 2.1.** *A nonempty open set $U$ in a metric space $X$ contains the closure of an open ball.*

*Proof.* Let $B(x, r)$ be an open ball in $U$. Clearly, $B(x, r/2) \subseteq B(x, r)$. Let $y$ be a limit point of $B(x, r/2)$. There exists $z \in B(x, r/2)$ such that $d(z, y) < r/2$. Hence

$$d(x, y) \le d(x, z) + d(z, y) < r/2 + r/2 = r,$$

so $y \in B(x, r)$. It follows that $\overline{B(x, r/2)} \subseteq B(x, r) \subseteq U$.                    □

Next we prove two forms of the main result. Both are known as the Baire category theorem.

**Theorem 2.16.** *Let $X_1, X_2, \ldots$ be a sequence of dense open sets in a complete metric space $(X, d)$. Then*

$$\bigcap_{n=1}^{\infty} X_n \ne \varnothing.$$

*Proof.* By Lemma 2.1, there is an open ball $B(x_1, \varepsilon_1)$ such that

$$\overline{B(x_1, \varepsilon_1)} \subseteq X_1.$$

Because $X_2$ is dense in $X$, the intersection $B(x_1, \varepsilon_1) \cap X_2$ is a nonempty open set, and therefore, by Lemma 2.1, there is an open ball $B(x_2, \varepsilon_2)$ such that

$$B(x_1, \varepsilon_1) \supseteq B(x_2, \varepsilon_2) \qquad \text{and} \qquad \overline{B(x_2, \varepsilon_2)} \subseteq X_2.$$

Clearly, we may assume that $0 < \varepsilon_2 < \varepsilon_1/2$. We construct recursively a nested sequence of open balls such that

$$B(x_1, \varepsilon_1) \supseteq B(x_2, \varepsilon_2) \supseteq \cdots \supset B(x_n, \varepsilon_n) \supseteq \cdots,$$

$$\overline{B(x_n, \varepsilon_n)} \subseteq X_n, \qquad \text{for all } n \in \mathbf{N},$$

and $0 < \varepsilon_n < \varepsilon_1/2^n$. By Theorem 2.9,

$$\bigcap_{n=1}^{\infty} X_n \supseteq \bigcap_{n=1}^{\infty} \overline{B(x_n, \varepsilon_n)} \neq \varnothing$$

(cf. Exercise 2.38), which is the desired result. $\qquad \qquad \square$

*Example 2.15.* Let $\mathbf{Q} = \{r_1, r_2, \ldots\}$ be an enumeration of the rational numbers. Define $X_n = (-\infty, r_n) \cup (r_n, \infty)$. Clearly, every set $X_n$ is open and dense in $\mathbf{R}$. It is not difficult to verify that the intersection $\bigcap_{n=1}^{\infty} X_k$ is the set $\mathbf{R} \backslash \mathbf{Q}$ of all irrational numbers in $\mathbf{R}$. This example hints at a generalization of Theorem 2.16 (cf. Exercise 2.53).

**Theorem 2.17.** *Let $(X, d)$ be a complete metric space. If $X = \bigcup_{n=1}^{\infty} F_n$, where $F_1, F_2, \ldots$ are closed sets, then at least one of these sets contains an open ball.*

*Proof.* Suppose that none of the closed sets $F_1, F_2, \ldots$ contains an open ball. Then each of the open sets $X \setminus F_1, X \setminus F_2, \ldots$ is dense in $X$ and

$$\bigcap_{n=1}^{\infty} (X \setminus F_n) = X \setminus \bigcup_{n=1}^{\infty} F_n = \varnothing,$$

which contradicts the result of Theorem 2.16. $\qquad \qquad \square$

*Example 2.16.* The set of real numbers $\mathbf{R}$ is not countable (cf. Example 2.14). Indeed, suppose that it is countable. Since $\mathbf{R} = \bigcup_{x \in \mathbf{R}} \{x\}$ and each set $\{x\}$ is closed, we have a contradiction with Theorem 2.17.

The metric space $\mathbf{Q}$ of rational numbers is a countable union of its singletons, $\mathbf{Q} = \bigcup_{x \in \mathbf{Q}} \{x\}$. By Theorem 2.17, $\mathbf{Q}$ is an incomplete space (cf. Example 2.13).

## 2.7 Compactness

**Definition 2.12.** A metric space $X$ is said to be *compact* if every sequence in $X$ has a convergent subsequence. A subset $Y$ of $X$ is called *compact* if it is a compact subspace of $X$.

Hence a subset $Y$ of a metric space $X$ is compact if every sequence in this subset has a subsequence that converges to a point in $Y$.

*Example 2.17.* In real analysis, the Bolzano–Weierstrass theorem asserts that a subset of $\mathbf{R}^n$ is compact if and only if it is closed and bounded.

**Theorem 2.18.** *A compact set in a metric space is closed and bounded.*

*Proof.* Let $E$ be a compact set in a metric space $X$. If $x$ is a point in the closure of $E$, then there is a sequence $(x_n)$ of points in $E$ that converges to $x$ in the space $X$. Because $E$ is compact, $x$ belongs to $E$. Hence $E$ is closed.

Suppose that the set $E$ is not bounded. Then for every $n \in \mathbf{N}$, there is a point $y_n \in E$ such that $d(y_n, a) > n$, where $a$ is a fixed point in $X$. Because every convergent sequence of points in $X$ is bounded, the sequence $(y_n)$ cannot contain a convergent subsequence. This contradicts our assumption that $E$ is compact. $\qquad\Box$

The converse of this theorem is false for general metric spaces, as the following example demonstrates.

*Example 2.18.* Let $X$ be an infinite set endowed with the discrete metric

$$d(x, y) = 1, \qquad \text{for all } x \neq y \text{ in } X.$$

The set $X$ is closed and bounded. Clearly, a sequence $(x_n)$ all of whose terms are distinct does not contain a convergent subsequence. (No subsequence of this sequence is Cauchy.) Hence, $X$ is not compact.

Let $Y$ be a subset of a metric space $X$. A family $\{U_i\}_{i \in J}$ of open sets in $X$ is said to be an *open covering* of $Y$ if $Y \subseteq \bigcup_{i \in J} U_i$. If $J' \subseteq J$ and $Y \subseteq \bigcup_{i \in J'} U_i$, then the family $\{U_i\}_{i \in J'}$ is called a *open subcovering* of $\{U_i\}_{i \in J}$.

**Lemma 2.2.** *Let $\{U_i\}_{i \in J}$ be an open covering of a compact space $X$. Then there exists $r > 0$ such that for every $x \in X$, the open ball $B(x, r)$ is contained in $U_i$ for some $i \in J$.*

*Proof.* Suppose to the contrary that for every $n \in \mathbf{N}$ there is $x_n \in X$ such that $B(x_n, 1/n) \not\subseteq U_i$ for all $i \in J$. Inasmuch as $X$ is compact, there is a subsequence $(x_{n_k})$ of the sequence $(x_n)$ converging to some $x \in X$. The point $x$ belongs to one of the sets $U_i$, say $x \in U_{i_0}$. Because $U_{i_0}$ is open, there exists $m$ such that $B(x, 1/m) \subseteq U_{i_0}$. Since $x_{n_k} \to x$, there exists $n_0 \geq 2m$ such that $x_{n_0} \in B(x, 1/2m)$. We have (cf. Exercise 2.13 (b)

$$B\left(x_{n_0}, \frac{1}{n_0}\right) \subseteq B\left(x_{n_0}, \frac{1}{2m}\right) \subseteq B\left(x, \frac{1}{m}\right) \subseteq U_{i_0},$$

which contradicts our assumption that $B(x_n, 1/n) \not\subseteq U_i$ for all $n \in \mathbf{N}$ and $i \in J$. $\qquad\Box$

**Theorem 2.19. (Borel–Lebesgue)** *A metric space $X$ is compact if and only if every open covering $\{U_i\}_{i \in J}$ of $X$ contains a finite subcovering.*

*Proof.* (Necessity.) Let $X$ be a compact space and $\{U_i\}_{i \in J}$ an open covering of $X$. By Lemma 2.2, there exists $r > 0$ such that for every $x \in X$, we have $B(x, r) \subseteq U_i$ for some $i \in J$. It suffices to prove that $X$ can be covered by a finite number of balls $B(x, r)$. If $B(x_1, r) = X$ for some $x_1 \in X$, we are done. Otherwise, choose $x_2 \in X \setminus B(x_1, r)$. If $B(x_1, r) \cup B(x_2, r) = X$, the proof is over. If by continuing this process we obtain $X$ on some step, the claim is proven. Otherwise, there exists a sequence $(x_n)$ of points of $X$ such that

$$x_{n+1} \notin B(x_1, r) \cup \cdots \cup B(x_n, r)$$

for every $n \in \mathbf{N}$. It is clear that $d(x_m, x_n) \geq r$ for all $m, n \in \mathbf{N}$. It follows that $(x_n)$ has no Cauchy subsequences and hence no convergent subsequences. This contradicts our assumption that $X$ is a compact space. Hence, $X$ can be covered by a finite number of sets $U_i$.

(Sufficiency.) Let $X$ be a metric space such that every open covering of $X$ contains a finite subcovering and let $(x_n)$ be a sequence of points in $X$. Suppose that $(x_n)$ does not have a convergent subsequence. Then for every point $x \in X$ there is an open ball $B(x, r_x)$ that contains no points of the sequence $(x_n)$ except possibly $x$ itself (cf. Exercise 2.28). The family $\{B(x, r_x)\}_{x \in X}$ is an open covering of $X$ and therefore contains a finite subcovering. Hence,

$$X = B(a_1, r_1) \cup \cdots \cup B(a_n, r_n),$$

for a finite set $A = \{a_1, \ldots, a_n\}$ in $X$. By the choice of open balls $B(x, r_x)$, we have $x_k \in A$ for all $k \in \mathbf{N}$, which contradicts our assumption that $(x_n)$ does not have a convergent subsequence (cf. Exercise 2.27). $\quad\square$

**Theorem 2.20.** *The image of a compact set under a continuous mapping is compact.*

*Proof.* Let $T : X \to Y$ be a continuous mapping and $A$ a compact subset of $X$. If $\{U_i\}_{i \in J}$ is an open covering of $T(A)$, then by Theorem 2.19, $\{T^{-1}(U_i)\}_{i \in J}$ is an open covering of $A$. Inasmuch as $A$ is compact, there is a subcovering $\{T^{-1}(U_i)\}_{i \in J'}$ of $A$, where $J'$ is a finite subset of $J$. Then $\{U_i\}_{i \in J'}$ is a finite subfamily of $\{U_i\}_{i \in J}$ that covers $T(A)$. Hence $T(A)$ is a compact subset of the space $Y$. $\quad\square$

**Theorem 2.21.** *If $X$ is a compact space, then every continuous function $f : X \to \mathbf{R}$ is bounded and attains its maximum value.*

*Proof.* By Theorem 2.20, the set $A = f(X) \subseteq \mathbf{R}$ is compact. If $A$ has no largest element, then the family

$$\{(-\infty, a) : a \in A\}$$

forms an open covering of $A$. Inasmuch as $A$ is compact, some finite subfamily

$$\{(-\infty, a_1), \ldots (-\infty, a_n)\}$$

covers $A$. If $a_k = \max\{a_1, \ldots, a_n\}$, then for all $1 \le i \le n$, we have $a_k \notin (-\infty, a_i)$, contrary to the fact that these intervals cover $A$.                                    $\square$

## 2.8 Topological Spaces

Many problems in functional analysis require the consideration of topological spaces, a more general structure than that of a metric space.

**Definition 2.13.** A *topological space* is a pair $(X, \mathcal{T})$, where $X$ is a nonempty set and $\mathcal{T}$ is a family of subsets of $X$, called *open* sets, possessing the following properties:

(a) The union of every collection of open sets is open.
(b) The intersection of every finite collection of open sets is open.
(c) $\varnothing \in \mathcal{T}$ and $X \in \mathcal{T}$.

Elements of the set $X$ are called *points* of the topological space $(X, \mathcal{T})$, and the family $\mathcal{T}$ is called a *topology on* $X$. For a point $x \in X$, a *neighborhood* of $x$ is a set containing an open set that contains $x$.

It is worth noting that property (c) of the definition follows from properties (a) and (b). Namely, the union of the empty collection of open sets is the empty set $\varnothing$, and the intersection of the empty collection is the set $X$ itself.

As in the case of metric spaces, it is customary to omit reference to the topology $\mathcal{T}$ in the notation $(X, \mathcal{T})$ and write "a topological space $X$" instead of "a topological space $(X, \mathcal{T})$."

Let $(X, \mathcal{T})$ be a topological space. If $Y$ is a subset of $X$, then the family

$$\mathcal{T}' = \{Y \cap U : U \in \mathcal{T}\}$$

is a topology on $Y$ (cf. Exercise 2.59). This topology is called the *subspace topology*.

An important example of a topological space is a metric space. Let $(X, d)$ be a metric space and $\mathcal{T}$ the family of all open sets in $X$. Note that $\mathcal{T}$ contains the empty set $\varnothing$ and the set $X$ itself. By Theorem 2.1, the pair $(X, \mathcal{T})$ is a topological space. In this case, the topology $\mathcal{T}$ is called the *metric topology* induced by the metric $d$.

*Example 2.19.* Let $X$ be a nonempty set. Define $\mathcal{T} = \{\varnothing, X\}$. It is clear that $\mathcal{T}$ is a topology on $X$. It is called the *trivial topology* on $X$.

*Example 2.20.* Let $X$ be a nonempty set and $\mathcal{T}$ the family of all subsets of $X$. Hence, $\mathcal{T}$ is the power set $2^X$. This topology is called the *discrete topology* on $X$. The discrete topology is induced by the discrete metric.

A topological space is said to be *Hausdorff* if every two distinct points in the space have disjoint neighborhoods.

*Example 2.21.* (a) Every metric space $(X, d)$ is Hausdorff. Indeed, if $x$ and $y$ are distinct points in $X$, then $B(x, d(x, y)/2) \cap B(y, d(x, y)/2) = \varnothing$.

(b) Let $X = \{a, b\}$ be a two-element set and $\mathcal{T} = \{\varnothing, \{a\}, X\}$. It is clear that $\mathcal{T}$ is a topology on $X$ and that the space $(X, \mathcal{T})$ is not Hausdorff.

*Example 2.22.* Let $(X, d)$ be a pseudometric space that is not a metric space, so $d(x, y) = 0$ for some distinct points $x$ and $y$ in $X$. Open balls and open sets in this space are defined exactly as in the case of metric spaces (cf. Definitions 2.3 and 2.4). It is not difficult to show that the space $X$ is not Hausdorff (cf. Exercise 2.58).

A *base* (or *basis*) of the topology $\mathcal{T}$ on a nonempty set $X$ is a family $\mathcal{B}$ of open subsets of $X$ such that every open subset of $X$ is the union of sets belonging to $\mathcal{B}$. If $\mathcal{B}$ is a base of the topology $\mathcal{T}$, we say that $\mathcal{B}$ *generates* $\mathcal{T}$.

Bases are useful because many topologies are most easily defined in terms of bases generating them. The following theorem (cf. Exercise 2.60) gives necessary and sufficient conditions for a family of subsets to generate a topology.

**Theorem 2.22.** *Let $\mathcal{B}$ be a family of subsets of a nonempty set $X$. Then $\mathcal{B}$ is a base of a topology on $X$ if and only if*

(a) $X = \bigcup_{B \in \mathcal{B}} B$;
(b) *if $B_1, B_2 \in \mathcal{B}$ and $x \in B_1 \cap B_2$, then there is a set $B$ in $\mathcal{B}$ such that $x \in B \subseteq B_1 \cap B_2$.*

*The unique topology that has $\mathcal{B}$ as its base consists of the unions of subfamilies of $\mathcal{B}$.*

*Example 2.23.* It is known from real analysis that every open subset of $\mathbf{R}$ is the (possibly empty) union of a family of open intervals. Hence, the open intervals form a base of the metric topology on $\mathbf{R}$. In fact, in every metric space $X$, the collection of open balls is a base of the metric topology on $X$.

**Definition 2.14.** Let $X$ and $Y$ be topological spaces. A mapping $T : X \to Y$ is said to be *continuous at a point* $x_0 \in X$ if to every neighborhood $U$ of the point $Tx_0$ there corresponds a neighborhood $V$ of point $x_0$ such that $T(V) \subseteq U$. The mapping $T$ is said to be *continuous* if it is continuous at every point of $X$.

**Definition 2.15.** Topological spaces $X$ and $Y$ are said to be *homeomorphic* if there is a bijection $T : X \to Y$ such that both mappings $T$ and $T^{-1}$ are continuous. A topological space is said to be *metrizable* if it is homeomorphic to a metric space.

If two metric spaces are isometric, then they are homeomorphic as topological spaces. Note that the converse is not true (cf. Exercise 2.61).

We have the following analogue of Theorem 2.11.

**Theorem 2.23.** *Let $X$ and $Y$ be topological spaces and $T$ a mapping from $X$ into $Y$. Then $T$ is continuous if and only if the inverse image under $T$ of every open set in $Y$ is an open set in $X$.*

*Proof.* (Necessity.) Let $T : X \to Y$ be a continuous mapping and $U$ an open set in $Y$. Then $U$ is a neighborhood of every point $Tx$ for $x \in T^{-1}(U)$. Because $T$ is continuous, every $x \in T^{-1}(U)$ has an open neighborhood $V$ such that $T(V) \subseteq U$, so $V \subseteq T^{-1}(U)$. Hence, $T^{-1}(U)$ is an open set.

(Sufficiency.) Suppose that the inverse image of every open set in $Y$ is an open set in $X$. For $x \in X$, a neighborhood of $Tx$ contains an open subset $U$ containing $Tx$. The set $V = T^{-1}(U)$ is an open neighborhood of $x$ such that $T(V) \subseteq U$. Hence $T$ is continuous at $x$. Because $x$ is an arbitrary point of $X$, the mapping $T$ is continuous. $\square$

**Definition 2.16.** If $(X_1, \mathcal{T}_1)$ and $(X_2, \mathcal{T}_2)$ are topological spaces, a topology on the Cartesian product $X_1 \times X_2$ is defined by taking as a base the collection of all sets of the form $U_1 \times U_2$, where $U_1 \in \mathcal{T}_1$ and $U_2 \in \mathcal{T}_2$. It can be verified (cf. Exercise 2.62) that this collection is indeed a base of a topology. This unique topology is called the *product topology* on $X_1 \times X_2$.

The result of Theorem 2.19 motivates the following definition.

**Definition 2.17.** A topological space $(X, \mathcal{T})$ is said to be *compact* if every covering of $X$ by open sets (that is, an open covering) contains a finite subcovering. A subset $Y$ of $X$ is said to be compact if $Y$ is a compact space in the subspace topology.

The proofs of the next two theorems can be taken verbatim from the proofs of Theorems 2.20 and 2.21.

**Theorem 2.24.** *Let $T : X \to X'$ be a continuous mapping of a topological space $(X, \mathcal{T})$ into a topological space $(X', \mathcal{T}')$. Then the image $T(Y)$ of a compact set $Y$ in $X$ is a compact set in $X'$.*

**Theorem 2.25.** *If $X$ is a compact space, then every continuous function $f : X \to \mathbf{R}$ is bounded and attains its maximum value.*

## Notes

Axioms **(M1)**–**(M3)** are motivated by classical Euclidean geometry, where in particular, it is proved that each side of a triangle is smaller than the sum of the other two sides, and each side is greater than the difference of the other two sides (see, for instance, Kiselev 2006, pp. 38–39).

There is a loose connection between the concept of a limit and that of a limit point of a subset. Let $E$ be a nonempty subset of a metric space and $x$ a limit point of $E$. For every $n \in \mathbf{N}$, there is a point $x_n \in E$ (distinct from $x$) such that $d(x_n, x) < 1/n$, so $x_n \to x$. Hence, a limit point of the set $E$ is the limit of a sequence of points in $E$. The converse is not true. For instance, if $E = X$, where $X$ is a discrete space, then $E$ has no limit points, whereas every point of $E$ is a limit of a constant sequence.

There is standard terminology associated with the Baire category theorems. A subset of a topological space is called *nowhere dense* (or *rare*) if its closure contains no interior points. Every countable union of nowhere dense sets is said to be of the *first category* (or *meager*). All other subsets are of *second category*. The standard Baire category theorem says that every complete metric space is of second category. This result was established by René-Louis Baire in his dissertation in 1899 for the spaces $\mathbf{R}^n$ and independently by William Fogg Osgood for the real line in 1897.

## Exercises

**2.1.** Prove that properties **M1–M3** of the metric $d$ are equivalent to the following conditions:

(1) $d(x, y) = 0$ if and only if $x = y$;
(2) $d(x, y) \leq d(x, z) + d(y, z)$,

for all $x, y, z \in X$.

**2.2.** Prove that the system of inequalities in (2.1) is equivalent to the system of three distinct triangle inequalities on three points.

**2.3.** (The quadrilateral inequality.) Prove the following generalization of the first inequality in (2.1). For every four points $x$, $y$, $u$, and $w$ in a metric space,

$$|d(x, y) - d(u, v)| \leq d(x, u) + d(y, v).$$

**2.4.** Prove inequality (2.2).

**2.5.** Let $d$ be a metric on a set $X$. Show that the following functions are also metrics on $X$:

(a) $\tilde{d} = \dfrac{d}{1 + d}$,
(b) $\tilde{d} = \ln(1 + d)$,
(c) $\tilde{d} = d^{\alpha}, \quad 0 < \alpha < 1$.

**2.6.** For a pseudometric space $(X, d)$, we define a binary relation $\sim$ on $X$ by

$$x \sim y \quad \text{if} \quad d(x, y) = 0, \qquad \text{for } x, y \in X.$$

Show that $\sim$ is an equivalence relation on $X$ and the quotient set $X/\sim$ (cf. Section 1.4) is a metric space with the distance function $\tilde{d}$ given by $\tilde{d}([x], [y]) = d(x, y)$.

**2.7.** Show that a seminorm $p$ on a vector space $X$ has the following properties:

(a) $p(0) = 0$,
(b) $p(-x) = p(x)$,
(c) $p(x) \geq 0$,

for all $x \in X$. Also show that the function $d(x, y) = p(x-y)$ is a pseudometric on $X$.

**2.8.** Let $d_1, \ldots, d_n$ be (pseudo)metrics on a set $X$ and $\lambda_1, \ldots, \lambda_n$ nonnegative real numbers such that at least one of them is positive. Show that

$$d = \lambda_1 d_1 + \cdots + \lambda_n d_n$$

is a (pseudo)metric on $X$.

**2.9.** Let $(d_1, \ldots, d_n, \ldots)$ be a sequence of (pseudo)metrics on a set $X$. Show that

$$\tilde{d} = \sum_{k=1}^{\infty} \frac{1}{2^k} \frac{d_k}{1 + d_k}$$

is a (pseudo)metric on $X$.

**2.10.** Let $y$ be a point in the ball $B(x, r)$. Show that there exists $r' > 0$ such that $B(y, r') \subseteq B(x, r)$. Conclude that an open ball is an open set.

**2.11.** Give an example of a metric space $(X, d)$ and open balls $B(x_1, r_1)$ and $B(x_2, r_2)$ in $X$ such that $B(x_2, r_2)$ is a proper subset of $B(x_1, r_1)$ and $r_2 > r_1$.

**2.12.** Suppose that $B(x, r_1)$ is a proper subset of $B(x, r_2)$ in a metric space $(X, d)$. Show that $r_1 < 2r_2$.

**2.13.** For points $x_1$ and $x_2$ in a metric space and positive numbers $r_1$ and $r_2$, prove that

(a) $d(x_1, x_2) \geq r_1 + r_2$ implies $B(x_1, r_1) \cap B(x_2, r_2) = \varnothing$.
(b) if $d(x_1, x_2) \leq r_1 - r_2$, then $B(x_2, r_2) \subseteq B(x_1, r_1)$.

Show that the converses of (a) and (b) do not necessarily hold.

**2.14.** Suppose that the intersection of two open balls in a metric space is not empty:

$$B(x, r_1) \cap B(y, r_2) \neq \varnothing.$$

Give an example of an open ball that is contained in this intersection.

**2.15.** Show that the empty set $\varnothing$ is an open subset of every metric space.

**2.16.** Prove that int $\mathbf{Q} = \varnothing$.

**2.17.** Show that a metric space consisting of more than one point has at least two open subsets different from $\varnothing$ and $X$.

**2.18.** Prove that in every metric space, the closure of an open ball is a subset of the closed ball with the same center and radius:

$$\overline{B(x,r)} \subseteq \bar{B}(x,r).$$

Give an example of a metric space and an open ball in it for which the above inclusion is proper.

**2.19.** Prove Theorem 2.4.

**2.20.** Let $A$ and $B$ be sets in a metric space. Show that
(a) $\bar{A} \subseteq \bar{B}$ if $A \subseteq B$.
(b) $\overline{A \cup B} = \bar{A} \cup \bar{B}$.
(c) $\overline{A \cap B} \subseteq \bar{A} \cap \bar{B}$.

Give an example of a proper inclusion in (c).

**2.21.** Let $X$ be a subspace of a metric space $Y$, and $Z$ a subset of $X$. Show that $Z$ is closed in $X$ if and only if there is a closed subset $F$ of $Y$ such that $Z = F \cap X$.

**2.22.** Show that an open set in a metric space $(X, d)$ is also open in the space $(X, \tilde{d})$, where $\tilde{d} = d/(1 + d)$ (cf. Exercise 2.5(a)). Is the converse true?

**2.23.** Show that the metric space $\mathbf{C}$ is separable.

**2.24.** Prove that a discrete metric space is separable if and only if it is countable.

**2.25.** Show that $x$ is an accumulation point of a subset $A$ of a metric space if and only if $x \in \overline{A \setminus \{x\}}$.

**2.26.** Prove that a subspace of a separable metric space is separable.

**2.27.** Let $(x_n)$ be a sequence of points in a metric space $X$ such that the set $\{x_n : n \in \mathbf{N}\}$ is finite. Show that $(x_n)$ contains a convergent subsequence.

**2.28.** Show that a point $x$ in a metric space $X$ is the limit of a subsequence of a sequence $(x_n)$ if and only if every neighborhood of $x$ contains infinitely many terms of $(x_n)$.

**2.29.** Let $(x_n)$ be a convergent sequence in a metric space $X$ and $x = \lim x_n$. Show that every subsequence $(x_{n_k})$ of $(x_n)$ converges to the same limit $x$.

**2.30.** If $(x_n)$ is Cauchy and has a convergent subsequence $(x_{n_k})$ with limit $x$, show that $(x_n)$ converges to the same limit $x$.

**2.31.** Show that a Cauchy sequence is bounded.

**2.32.** Suppose that $(x_n)$ is a Cauchy sequence in a metric space $(X, d)$. Show that there is a subsequence $(x_{n_k})$ of $(x_n)$ such that

$$d(x_{n_k}, x_{n_m}) < 1/2^k, \qquad \text{for all } m > k.$$

**2.33.** Show that a metric space is a singleton if and only if every bounded sequence is convergent.

**2.34.** Show that a sequence $(x_n)$ in a discrete metric space $X$ converges to $x \in X$ if and only if there is $N \in \mathbf{N}$ such that $x_n = x$ for all $n > N$.

**2.35.** (a) Show that the sequence $(1/n)$ diverges in the metric space $(0, 1)$ with the distance function inherited from $\mathbf{R}$.

(b) Show that this sequence is a Cauchy sequence in $(0, 1)$.

**2.36.** Let $(c_n)$ be a sequence of positive rational numbers such that $c_{k+1} < c_k$ for all $k \in \mathbf{N}$. Show that $\sum_{k=1}^{\infty}(-1)^{k-1}c_k = r$ implies

$$0 < \left| r - \sum_{k=1}^{n}(-1)^{k-1}c_k \right| < c_{n+1}, \qquad \text{for all } n \in \mathbf{N}.$$

**2.37.** Let $(X, d)$ be a metric space, where $X$ is a finite set. Show that this space is complete.

**2.38.** Let $B(x, r)$ be an open ball in a metric space $(X, d)$. Show that $\overline{B(x, r)}$ is a bounded set of diameter not greater than $2r$.

**2.39.** Let $A$ be a bounded subset of a metric space. Show that

$$\text{diam}(\overline{A}) = \text{diam}(A).$$

**2.40.** Show that the condition $\text{diam}(A_n) \to 0$ in Theorem 2.9 is essential. (Give an example of a complete metric space and a nested family of bounded closed sets in it with empty intersection.)

**2.41.** Prove the converse of Theorem 2.9 (Royden and Fitzpatrick 2010, Section 9.4).

**2.42.** Show that a mapping from a discrete metric space into a metric space is always continuous.

**2.43.** Let $a$ be a point in a metric space $(X, d)$. Show that the function $f(x) = d(x, a)$ is uniformly continuous on $X$, that is, for a given $\varepsilon > 0$ there exists $\delta > 0$ such that

$$|f(x) - f(y)| < \varepsilon,$$

for all $x, y \in X$ such that $d(x, y) < \delta$.

**2.44.** For a nonempty subset $A$ of a metric space $(X, d)$ and a point $x \in X$, let

$$\text{dist}(x, A) = \inf\{d(x, a) : a \in A\}.$$

Show that the function $f(x) = \text{dist}(x, A)$ is continuous.

**2.45.** Prove that a subset $U$ of a metric space $X$ is open if and only if there is a continuous function $f : X \to \mathbf{R}$ such that $U = \{x \in X : f(x) > 0\}$.

**2.46.** If $T : X \to Y$ is an isometry from a complete metric space $X$ into a metric space $Y$, then $T(X)$ is a complete subspace of $Y$.

**2.47.** Describe the completion of a discrete metric space.

**2.48.** If $X$ and $Y$ are isometric metric spaces and $X$ is complete, show that $Y$ is complete.

**2.49.** Let $(X, d)$ be a metric space and let $\tilde{d} = d/(1+d)$ (cf. Exercise 2.5(a)). Show that $(X, \tilde{d})$ is complete if and only if $(X, d)$ is complete.

**2.50.** Define $d(x, y) = |\tan^{-1} x - \tan^{-1} y|$ on $\mathbf{R}$. Show that $(\mathbf{R}, d)$ is an incomplete metric space and find its completion.

**2.51.** Let $X$ be the set of nonempty open intervals in $\mathbf{R}$ and let $d$ be defined by

$$d((x, y), (u, v)) = |u - x| + |v - y|.$$

(Here $(x, y)$ and $(u, v)$ are open intervals in $\mathbf{R}$.) Show that $(X, d)$ is an incomplete metric space and find its completion.

**2.52.** Give an example of an incomplete metric space $X$ and a sequence $(X_n)$ of open dense subsets of $X$ with empty intersection.

**2.53.** Let $(X_n)$ be a sequence of dense open sets in a complete metric space. Prove that the intersection of the sets in this sequence is dense in $X$.

**2.54.** A function $f : \mathbf{R} \to \mathbf{R}$ is said to be *lower semicontinuous* on $\mathbf{R}$ if the set $\{x : f(x) > \alpha\}$ is open for every $\alpha \in \mathbf{R}$. Let $f$ be a lower semicontinuous function on $\mathbf{R}$. Show that for every open set $U \subseteq \mathbf{R}$ there is an open subset $V \subseteq U$ on which $f$ is bounded. (Hint: use the Baire category theorem.)

**2.55.** Show that the set of irrational numbers in $[0, 1]$ cannot be represented as a countable union of closed sets. (Hint: Use the Baire category theorem.)

**2.56.** Prove that a set is dense in a metric space if and only if it has nonempty intersection with every open ball.

**2.57.** A topological space $(X, \mathcal{T})$ is said to be a $T_1$ space if for every two distinct points $x, y \in X$ there is an open set $U \in \mathcal{T}$ such that $x \in U$, $y \notin U$. Prove that a topological space is a $T_1$ space if and only if every singleton in $X$ is a closed set.

**2.58.** Let $(X, d)$ be a pseudometric space such that $d(x, y) = 0$ for some pair of distinct points $x$, $y$ in $X$. Show that there are no disjoint neighborhoods of $x$ and $y$ in the topological space $X$.

**2.59.** Let $(X, \mathcal{T})$ be a topological space and $X'$ a nonempty subset of $X$. Show that
$$\mathcal{T}' = \{U \cap X' : U \in \mathcal{T}\}$$
is a topology on $X'$.

**2.60.** Prove Theorem 2.22.

**2.61.** Give an example of two homeomorphic metric spaces that are not isometric.

**2.62.** Let $(X_1, \mathcal{T}_1)$ and $(X_2, \mathcal{T}_2)$ be topological spaces. Show that the set
$$\{U_1 \times U_2 : U_1 \in \mathcal{T}_1, \ U_2 \in \mathcal{T}_2\}$$
is a base of a topology on $X_1 \times X_2$. (Hint: Use Theorem 2.22.)

**2.63.** Show that a topology $\mathcal{T}$ on a set $X$ is discrete if and only if every mapping from $(X, \mathcal{T})$ into any topological space is continuous.

# Chapter 3
# Special Spaces

Topological, metric, and normed vector spaces are abstract spaces that together with operators on these spaces are major objects of study in functional analysis. These spaces are briefly introduced in Section 3.1. However, many concrete spaces not only serve as examples (and counterexamples) in the theory, but themselves are of significant theoretical importance. In the main body of this chapter (Sections 3.2–3.7) some of these spaces are described, and their main topological and metric properties are established.

## 3.1 Metric Vector Spaces

All spaces in functional analysis are vector spaces that are also topological spaces. In the following definition, the algebraic structure of a vector space is "blended" with a topology on the space.

**Definition 3.1.** A *topological vector space* is a vector space $X$ over the field **F** that is endowed with a topology such that vector addition and scalar multiplication are continuous functions.

Continuity of vector addition means that for every neighborhood $U$ of the vector $x + y$ there are neighborhoods $U_x$ and $U_y$ of vectors $x$ and $y$, respectively, such that

$$U_x + U_y \subseteq U.$$

Scalar multiplication is continuous if for every neighborhood $U$ of the vector $\lambda x$ ($\lambda \in \mathbf{F}$) there exist a neighborhood $V_\lambda$ of the scalar $\lambda$ and a neighborhood $U_x$ of the vector $x$ such that

$$V_\lambda U_x \subseteq U.$$

If a topological vector space is a metric (or pseudometric) space, that is, its topology is the topology of a metric (or pseudometric) space, then we call

© Springer International Publishing AG, part of Springer Nature 2018
S. Ovchinnikov, *Functional Analysis*, Universitext,
https://doi.org/10.1007/978-3-319-91512-8_3

it a *metric vector space*. In this case, continuity of vector addition and scalar multiplication can be expressed in terms of convergent sequences. Namely (cf. Exercise 3.1), a pair $(X, d)$ is a metric vector space if for sequences $(x_n)$ and $(y_n)$ of vectors in $X$ and sequences $(\lambda_n)$ of scalars in $\mathbf{F}$, these conditions hold:

(a)   $x_n \to x, \, y_n \to y$   imply   $(x_n + y_n) \to (x + y)$,
(b)   $\lambda_n \to \lambda, \, x_n \to x$   imply   $\lambda_n x_n \to \lambda x$.

If a metric vector space $(X, d)$ is complete, then it is called a *Fréchet space*.

We reintroduce below the concepts that were introduced in passing in Chapter 2.

**Definition 3.2.** Let $X$ be a vector space over the field $\mathbf{F}$. A function $p : X \to \mathbf{R}$ is said to be a *seminorm on $X$* if

(a)   $p(\lambda x) = |\lambda| p(x)$,                 *homogeneity*
(b)   $p(x + y) \le p(x) + p(y)$,         *triangle inequality*

for all $\lambda \in \mathbf{F}$ and $x, y \in X$. The pair $(X, p)$ is called a *seminormed space*.

A *norm* on $X$ is a real-valued function on $X$, whose value at $x \in X$ is denoted by $\|x\|$, with the properties

(a)   $\|\lambda x\| = |\lambda| \|x\|$,                 *homogeneity*
(b)   $\|x + y\| \le \|x\| + \|y\|$,         *triangle inequality*
(c)   $\|x\| = 0$   implies   $x = 0$,

for all $\lambda \in \mathbf{F}$ and $x, y \in X$. The pair $(X, \| \cdot \|)$ is said to be a *normed space*. A *Banach space* is a normed space that is a complete metric space.

Note that norms (and seminorms) are nonnegative functions (cf. Exercise 3.2).

Clearly a norm is also a seminorm. It can be easily verified that the functions $p(x - y)$ and $d(x, y) = \|x - y\|$ are a pseudometric and metric on $X$, respectively. Convergence in a normed space is defined by the metric $d$, that is, $x_n \to x$ in $(X, \| \cdot \|)$ if $\|x_n - x\| \to 0$ as $n \to \infty$.

Some basic properties of seminorms and norms are found in Exercises 2.7 and 3.2.

**Theorem 3.1.** *Normed and seminormed spaces are metric vector spaces.*

*Proof.* It suffices to consider the case of a seminormed space $(X, p)$ with the distance function $d(x, y) = p(x - y)$. Note that $p$ is a nonnegative function on $X$ (cf. Exercise 2.7).

Suppose that $x_n \to x$ and $y_n \to y$ in $(X, p)$, so $p(x_n - x) \to 0$ and $p(y_n - y) \to 0$. Then

$$0 \le p((x_n + y_n) - (x + y)) = p((x_n - x) + (y_n - y))$$
$$\le p(x_n - x) + p(y_n - y) \to 0.$$

Hence $(x_n + y_n) \to (x + y)$.

Now suppose that $\lambda_n \to \lambda$ in $\mathbf{F}$ and $x_n \to x$ in $(X, p)$, so $|\lambda_n - \lambda| \to 0$ and $p(x_n - x) \to 0$. Because $\lambda_n \to \lambda$, the sequence $(\lambda_n)$ is bounded. We have

$$
\begin{aligned}
0 \leq p(\lambda_n x_n - \lambda x) &= p(\lambda_n x_n - \lambda_n x + \lambda_n x - \lambda x) \\
&= p((\lambda_n(x_n - x) + (\lambda_n - \lambda)x) \\
&\leq |\lambda_n| p(x_n - x) + |\lambda_n - \lambda| p(x) \to 0.
\end{aligned}
$$

It follows that $\lambda_n x_n \to \lambda x$.                                                                         □

Note that a normed space is Hausdorff (cf. Example 2.21).

**Definition 3.3.** A *normed subspace* of a normed space $X$ is a vector subspace $Y$ of $X$ with the norm obtained by restricting the norm on $X$ to $Y$.

Abstract normed spaces are players in the theory presented in Chapter 4. Many particular spaces in functional analysis are metric vector spaces of functions. This means that every specific space is a vector space of $\mathbf{F}$-valued functions on some set. This vector space is endowed with a norm, seminorm, or metric that makes it a metric vector space. $\mathbf{F}$-valued functions on the set $\{1, 2, \ldots, n\}$ are $n$-dimensional vectors. The set of all these functions forms the vector space $\mathbf{F}^n$. Functions on the set of natural numbers $\mathbf{N}$ are sequences. Vectors in *sequence spaces* are sequences with terms in $\mathbf{F}$. In the rest of this chapter, we describe some classes of these and other special spaces in functional analysis and establish their completeness and separability properties.

## 3.2 Finite-Dimensional $\ell_p$ Spaces

Recall that vectors in the $n$-dimensional vector space $\mathbf{F}^n$ are $n$-tuples with terms in $\mathbf{F}$:

$$
\mathbf{F}^n = \{(x_1, \ldots, x_n) : x_i \in \mathbf{F}, \ 1 \leq i \leq n\}.
$$

In this section, we consider families of norms and metrics on $\mathbf{F}^n$ parameterized by the extended real variable $p \in (0, \infty) \cup \{\infty\} = (0, \infty]$.

For $p \in [1, \infty)$, we define

$$
\|x\|_p = \left( \sum_{i=1}^{n} |x_i|^p \right)^{1/p}, \qquad \text{for } x = (x_1, \ldots, x_n) \in \mathbf{F}^n. \tag{3.1}
$$

This function on $\mathbf{F}^n$ is a norm. Indeed, properties (a) and (c) of a norm in Definition 3.2 are trivially satisfied, and the triangle inequality for $\|\cdot\|_p$ is Minkowski's inequality (1.9). The normed space $(\mathbf{F}^n, \|\cdot\|_p)$ is denoted by $\ell_p^n$.

The real space $\ell_2^n$ is $n$-dimensional Euclidean space (often denoted by $\mathbf{R}^n$) with the metric defined by

$$d(x,y) = \|x - y\|_2 = \sqrt{\sum_{i=1}^{n} |x_i - y_i|^2}, \qquad x, y \in \mathbf{R}^n.$$

For $p = \infty$, the function

$$\|x\|_\infty = \max\{|x_i| : 1 \le i \le n\}, \qquad \text{for } x = (x_1, \dots, x_n) \in \mathbf{F}^n$$

is a norm on $\mathbf{F}^n$. Again, properties (a) and (c) of a norm in Definition 3.2 are trivially satisfied. The triangle inequality for $\|\cdot\|_\infty$ follows from the triangle inequalities

$$|x_i + y_i| \le |x_i| + |y_i|, \qquad 1 \le i \le n,$$

for the absolute value function on $\mathbf{F}$ (cf. Exercise 3.7). The resulting normed space is denoted by $\ell_\infty^n$ (cf. Exercise 3.8).

For $0 < p < 1$, the function $\|\cdot\|_p$ in (3.1) is not a norm (cf. Exercise 1.9). In this case, we define a metric on $\mathbf{F}^n$ by

$$d_p(x,y) = \sum_{i=1}^{n} |x_i - y_i|^p, \tag{3.2}$$

for $x = (x_1, \dots, x_n)$ and $y = (y_1, \dots, y_n)$ in $\mathbf{F}^n$. The triangle inequality for $d$ follows immediately from (1.13). Clearly, $d_p(x,y) = 0$ if and only if $x = y$, and $d_p(x,y) = d_p(y,x)$ for all $x, y \in \mathbf{F}^n$. Hence $d_p$ is indeed a metric on $\mathbf{F}^n$. We denote the metric space $(\mathbf{F}^n, d)$ by the same symbol $\ell_p^n$ as for $p \ge 1$.

The following two properties of the metric $d_p$ are essential in the proof of Theorem 3.2:

$$d_p(x + z, y + z) = d_p(x,y) \qquad \text{and} \qquad d_p(\lambda x, \lambda y) = |\lambda|^p d_p(x,y),$$

for all $x, y, z \in \ell_p^n$ and $\lambda \in \mathbf{F}$. The first property is known as *translation invariance* of the distance function.

**Theorem 3.2.** *For $p \in (0, \infty]$ and every $n \in \mathbf{N}$, the space $\ell_p^n$ is a metric vector space.*

*Proof.* It suffices to consider the case $p \in (0,1)$, because for $p \in [1, \infty]$ the space $\ell_p^n$ is a normed space, and the result follows from Theorem 3.1.

Suppose that $0 < p < 1$ and $x^{(k)} \to x$, $y^{(k)} \to y$ in $\ell_p^n$ as $k \to \infty$. By the translation-invariance property,

$$d_p(x^{(k)} - x, 0) = d_p(x^{(k)}, x) \to 0$$

and

$$d_p(y^{(k)} - y, 0) = d_p(y^{(k)}, y) \to 0.$$

By applying the same property and the triangle inequality, we obtain

$$0 \le d_p(x^{(k)} + y^{(k)}, x + y) = d_p(x^{(k)} - x, y - y^{(k)})$$
$$\le d_p(x^{(k)} - x, 0) + d_p(0, y - y^{(k)}) \to 0.$$

Therefore, $(x^{(k)} + y^{(k)}) \to (x + y)$, so vector addition is a continuous function.

Now we prove that scalar multiplication is a continuous function. Let $\lambda^{(k)} \to \lambda$ in $\mathbf{F}$ and $x^{(k)} \to x$ in $\ell_p^n$ as $k \to \infty$. We have

$$0 \le d_p(\lambda^{(k)} x^{(k)}, \lambda x) \le d_p(\lambda^{(k)} x^{(k)}, \lambda^{(k)} x) + d_p(\lambda^{(k)} x, \lambda x)$$
$$= d_p(0, \lambda^{(k)}(x - x^{(k)})) + d_p((\lambda^{(k)} - \lambda)x, 0)$$
$$= |\lambda^{(k)}|^p d_p(0, x - x^{(k)}) + |\lambda^{(k)} - \lambda|^p d_p(x, 0) \to 0.$$

(Note that $(\lambda^{(k)})$ is a bounded sequence, since it is convergent.) Hence $\lambda^{(k)} x^{(k)} \to \lambda x$. $\qquad\square$

**Theorem 3.3.** *All spaces $\ell_p^n$, $n \in \mathbf{N}$ and $p \in (0, \infty]$, are complete and separable.*

*Proof.* We establish the assertion of the theorem for $p = 2$, leaving the remaining cases for the reader (cf. Exercise 3.9). Let

$$x^{(k)} = (x_1^{(k)}, \dots, x_n^{(k)}), \qquad k = 1, 2, \dots,$$

be a Cauchy sequence in $\ell_2^n$. Then for every $\varepsilon > 0$, there exists $N \in \mathbf{N}$ such that

$$\|x^{(k)} - x^{(m)}\|_2 = \left( \sum_{i=1}^n |x_i^{(k)} - x_i^{(m)}|^2 \right)^{1/2} < \varepsilon, \qquad \text{for } k, m > N.$$

It follows that for $i = 1, \dots, n$,

$$|x_i^{(k)} - x_i^{(m)}| < \varepsilon \qquad \text{for } k, m > N.$$

Hence for a fixed $1 \le i \le n$, the sequence $(x_i^{(k)})$ is a Cauchy sequence in $\mathbf{F}$ and therefore converges to some $x_i \in \mathbf{F}$. Let $x = (x_1, \dots, x_n)$. We have

$$\|x - x^{(k)}\|_2 = \left( \sum_{i=1}^n |x_i - x_i^{(k)}|^2 \right)^{1/2}.$$

Inasmuch as $x_i^{(k)} \to x_i$ as $k \to \infty$, the right-hand side of the above equality converges to zero. Hence $x^{(k)} \to x$ in $\ell_2^n$, so the space $\ell_2^n$ is complete.

To show that $\ell_2^n$ is a separable metric space, we consider the set $M$ consisting of the sequences in $\ell_2^n$ with rational coordinates. (A complex number is said to be rational if its real and imaginary parts are rational numbers.) Let $x = (x_1, \dots, x_n)$ be a vector in $\ell_2^n$. Because the rational numbers are dense in $\mathbf{F}$, for every $\varepsilon > 0$ there exists $y = (y_1, \dots, y_n) \in M$ such that

$$|x_i - y_i| < \varepsilon/\sqrt{n}, \qquad \text{for } i = 1, \ldots, n.$$

Then

$$\|x - y\|_2 = \left( \sum_{i=1}^{n} |x_i - y_i|^2 \right)^{1/2} < \varepsilon.$$

It follows that the countable set $M$ is dense in $\ell_2^p$. Hence $\ell_2^p$ is separable. $\square$

**Definition 3.4.** Topological vector spaces $X$ and $Y$ are said to be *isomorphic* if there is a bijection $T : X \to Y$ such that

(a)    $T$ is an isomorphism of vector spaces $X$ and $Y$.
(b)    $T$ is a homeomorphism of topological spaces $X$ and $Y$.

Note that the relation of being isomorphic is transitive: if $X$ is isomorphic to $Y$ and $Y$ is isomorphic to $Z$, then $X$ is isomorphic to $Z$. This follows from the transitivity property of vector space isomorphisms and topological space homeomorphisms.

**Theorem 3.4.** *For every $n \in \mathbf{N}$, the spaces $\ell_p^n$ ($p \in (0, \infty]$) are isomorphic.*

*Proof.* By the transitivity property, it suffices to show that every space $\ell_p^n$ for $0 < p < \infty$ is isomorphic to $\ell_\infty^n$. Let $T$ be the identity mapping of the space $\mathbf{F}^n$ onto itself. Clearly, $T$ is a vector space isomorphism. To prove that $T$ is an isomorphism of $\ell_p^n$ onto $\ell_\infty^n$, we show that for every given $r > 0$, the ball $B_p(0, r)$ in the space $\ell_p^n$ contains an open ball $B_\infty(0, r')$ in the space $\ell_\infty^n$ and conversely (cf. Exercise 3.10).

Let $B_p(0, r)$ be an open ball in $\ell_p^n$. We consider two cases.

Case $0 < p < 1$. We set $r' = (r/n)^{1/p}$ and show that $B_\infty(0, r') \subseteq B_p(0, r)$. If $x \in B_\infty(0, r')$, then $\max\{|x_i| : 1 \le i \le n\} < (r/n)^{1/p}$, so $|x_i| < (r/n)^{1/p}$ for all $1 \le i \le n$. It follows that

$$\sum_{i=1}^{n} |x_i|^p < \sum_{i=1}^{n} \frac{r}{n} = r.$$

Hence $x \in B_p(0, r)$.

Case $1 \le p < \infty$. We set $r' = r/n^{1/p}$ and show that $B_\infty(0, r') \subseteq B_p(0, r)$. If $x \in B_\infty(0, r')$, then $\max\{|x_i| : 1 \le i \le n\} < r/n^{1/p}$, so $|x_i| < r/n^{1/p}$ for all $1 \le i \le n$. It follows that

$$\left( \sum_{i=1}^{n} |x_i|^p \right)^{1/p} < \left( \sum_{i=1}^{n} \frac{r^p}{n} \right)^{1/p} = r.$$

Hence $x \in B_p(0, r)$.

Now let $B_\infty(0, r)$ be an open ball in $\ell_\infty^n$.

For $1 \le p < \infty$, we have $B_p(0, r) \subseteq B_\infty(0, r)$, because

$$\left( \sum_{i=1}^{n} |x_i|^p \right)^{1/p} < r \quad \text{implies} \quad |x_i| < r, \quad \text{for all } 1 \le i \le n.$$

Furthermore, for $0 < p < 1$, we have $B_p(0, r^p) \subseteq B_\infty(0, r)$, because

$$\left( \sum_{i=1}^{n} |x_i|^p \right) < r^p \quad \text{implies} \quad |x_i| < r, \quad \text{for all } 1 \le i \le n.$$

The result follows. □

Later (cf. Theorem 4.12), we will prove that every two $n$-dimensional normed spaces are isomorphic.

## 3.3 Sequence Spaces $\ell_p$

The $\ell_p$ spaces are instances of sequence spaces, that is, vector spaces of **F**-valued functions on the set of natural numbers **N**. As in the previous section, these spaces are parameterized by the parameter $p \in (0, \infty]$.

For $p \in (0, \infty)$, vectors in $\ell_p$ are sequences $(x_n)$ for which the series

$$\sum_{i=1}^{\infty} |x_i|^p$$

converges. Clearly, $\ell_p$ is closed under scalar multiplication. By inequalities (1.12) and (1.13), it is closed under vector addition. Hence $\ell_p$ is a vector space.

For $p \in [1, \infty)$, we define

$$\|x\|_p = \left( \sum_{i=1}^{\infty} |x_i|^p \right)^{1/p}, \qquad \text{for } x = (x_1, x_2, \dots) \in \ell_p, \tag{3.3}$$

and for $0 < p < 1$,

$$d_p(x, y) = \sum_{i=1}^{\infty} |x_i - y_i|^p, \tag{3.4}$$

for $x = (x_1, x_2, \dots)$ and $y = (y_1, y_2, \dots)$ in $\ell_p$. As in the case of spaces $\ell_p^n$, the space $\ell_p$ is a normed space if $p \ge 1$ and a metric vector space if $0 < p < 1$.

The space $\ell_2$ is a prototypical example of a separable Hilbert space (cf. Chapter 7).

For $p = \infty$, the space $\ell_\infty$ is the vector space of all bounded sequences $(x_n)$ endowed with the *supremum norm* (also called the *sup-norm*)

$$\|x\|_\infty = \sup\{|x_i| : i \in \mathbf{N}\}\} \tag{3.5}$$

(cf. Exercise 3.12).

In the rest of this section, we establish the completeness and separability properties of the normed spaces $\ell_p$ $(p \in [1, \infty])$. The arguments employed in the proofs are typical in analysis and can be used to establish these properties for other normed and metric vector spaces (cf. Exercises 3.13 and 3.14).

**Theorem 3.5.** *The space $\ell_p$ $(p \in [1, \infty))$ is complete.*

*Proof.* Let $(x^{(n)})$ be a Cauchy sequence in $\ell_p$, where $x^{(n)} = (x_1^{(n)}, x_2^{(n)}, \ldots)$. Then for every $\varepsilon > 0$ there exists $N \in \mathbf{N}$ such that for all $m, n > N$,

$$\|x^{(m)} - x^{(n)}\|_p = \left( \sum_{i=1}^{\infty} |x_i^{(m)} - x_i^{(n)}|^p \right)^{1/p} < \varepsilon. \tag{3.6}$$

Hence for every $i$,

$$|x_i^{(m)} - x_i^{(n)}| < \varepsilon, \qquad \text{for all } m, n > N.$$

It follows that for every $i$, the sequence $(x_i^{(n)})$ is a Cauchy sequence in $\mathbf{F}$ and therefore converges to some limit $x_i$ as $n \to \infty$. We define $x = (x_1, x_2, \ldots)$ and show that $x \in \ell_p$ and $x^{(m)} \to x$ as $m \to \infty$.

For $k \in \mathbf{N}$, we obtain from (3.6),

$$\sum_{i=1}^{k} |x_i^{(m)} - x_i^{(n)}|^p < \varepsilon^p, \qquad \text{for all } m, n > N.$$

By taking repeated limits as $n \to \infty$ first and then $k \to \infty$, we obtain

$$\sum_{i=1}^{\infty} |x_i^{(m)} - x_i|^p \leq \varepsilon^p, \qquad \text{for all } m > N, \tag{3.7}$$

which means that $(x^{(m)} - x) \in \ell_p$. Hence

$$x = x^{(m)} + (x - x^{(m)}) \in \ell_p.$$

Finally, by (3.7), $\|x^{(m)} - x\|_p \leq \varepsilon$ for $m > N$, so $x^{(m)} \to x$ in $\ell_p$. Hence $\ell_p$ is a complete metric space. $\qquad \square$

We use similar steps in the proof of the following theorem.

**Theorem 3.6.** *The space $\ell_\infty$ is complete.*

*Proof.* Let $(x^{(n)})$ be a Cauchy sequence in $\ell_p$, where $x^{(n)} = (x_1^{(n)}, x_2^{(n)}, \ldots)$. Then for every $\varepsilon > 0$, there exists $N \in \mathbf{N}$ such that for all $m, n > N$,

$$\|x^{(m)} - x^{(n)}\|_\infty = \sup_{i \in \mathbf{N}} |x_i^{(m)} - x_i^{(n)}| < \varepsilon. \tag{3.8}$$

Hence for every $i \in \mathbf{N}$,

$$\left|x_i^{(m)} - x_i^{(n)}\right| < \varepsilon, \qquad \text{for all } m, n > N.$$

It follows that for every $i$, the sequence $(x_i^{(n)})$ is a Cauchy sequence in $\mathbf{F}$ and therefore converges to some limit $x_i$ as $n \to \infty$. We define $x = (x_1, x_2, \ldots)$ and show that $x \in \ell_\infty$ and $x^{(m)} \to x$ as $m \to \infty$.

By taking $n \to \infty$ in the previous displayed inequality, we obtain

$$\left|x_i^{(m)} - x_i\right| \leq \varepsilon, \qquad \text{for all } m > N \text{ and } i \in \mathbf{N}. \tag{3.9}$$

Hence for some $m$,

$$|x_i| = |x_i^{(m)} + (x_i - x_i^{(m)})| \leq |x_i^{(m)}| + \varepsilon \qquad \text{for all } i \in \mathbf{N}.$$

Inasmuch as the sequence $x^{(m)}$ is bounded, so is the sequence $x$. Therefore, $x \in \ell_\infty$. By (3.9), $\|x^{(m)} - x\|_\infty \leq \varepsilon$ for all $m > N$. Hence $x^{(m)} \to x$ in $\ell_\infty$, so $\ell_\infty$ is complete. $\qquad\square$

Theorems 3.5 and 3.6 claim that all normed (so $p \in [1, \infty]$) spaces $\ell_p$ are Banach spaces. Note that this also holds for finite-dimensional normed spaces (cf. Exercise 3.13).

Now we show that $\ell_p$ is a separable space for $p \in [1, \infty)$ and that $\ell_\infty$ is not separable. The proof for other values of $p$ is similar (cf. Exercise 3.14).

**Theorem 3.7.** *If $p \in [1, \infty)$, then the space $\ell_p$ is separable. The space $\ell_\infty$ is not separable.*

*Proof.* Let $M$ be the set of all sequences in $\ell_p$ with rational coordinates. Because $M$ is a countable set, it suffices to show that $\overline{M} = \ell_p$, that is, for a given $\varepsilon > 0$ and $x \in \ell_p$, there exists $y \in M$ such that $\|x - y\|_p < \varepsilon$.

Inasmuch as $x = (x_1, x_2, \ldots) \in \ell_p$, the series $\sum_{i=1}^{\infty} |x_i|^p$ converges. Therefore, there exists $n$ such that

$$\sum_{i=n+1}^{\infty} |x_i|^p < \frac{1}{2}\varepsilon^p.$$

Because rational numbers are dense in $\mathbf{F}$, we can choose rational numbers $y_i$ for $1 \leq i \leq n$ such that

$$\sum_{i=1}^{n} |x_i - y_i|^p < \frac{1}{2}\varepsilon^p.$$

Then for $y = (y_1, \ldots, y_n, 0, 0, \ldots)$, we have

$$\|x - y\|_p = \left(\sum_{i=1}^{n} |x_i - y_i|^p + \sum_{i=n+1}^{\infty} |x_i|^p\right)^{1/p} < \varepsilon,$$

which is the desired result.

Now we show that the space $\ell_\infty$ is not separable. Let

$$M = \{x = (x_1, x_2, \ldots) \in \ell_\infty : x_k \in \{0, 1\}, \ k \in \mathbf{N}\}.$$

The set is uncountable. Indeed, for $x \in M$ let $N_x = \{k \in \mathbf{N} : x_k = 1\}$. Then $M$ is in one-to-one correspondence with the power set $2^{\mathbf{N}}$ of all subsets of the set of natural numbers $\mathbf{N}$. It is known that $2^{\mathbf{N}}$ is not a countable set.

For $x \neq y$ in $M$, we have $\|x - y\|_\infty = 1$, so open balls centered at vectors in $M$ and of radius $1/2$ are disjoint. Because there are uncountably many of these balls, every countable subset of $\ell_\infty$ is disjoint from one of these balls. Hence a countable set cannot be dense in $\ell_\infty$, so the space is not separable.                                                                          $\square$

We conclude this section by introducing a relation between spaces $\ell_p$. The proof of the following theorem is left as an exercise (cf. Exercise 3.22).

**Theorem 3.8.** *If $p < q$, then the vector space $\ell_p$ is a proper subspace of the space $\ell_q$. Furthermore, every space $\ell_p$ is a proper subspace of the space $\ell_\infty$.*

Note that in Theorem 3.8, all spaces are vector spaces. In fact, $\ell_p$ is not a normed subspace (cf. Definition 3.3) of any $\ell_q$ $(q \neq p)$ (cf. Exercise 3.16).

## 3.4 Sequence Spaces $c$, $c_0$, and $c_{00}$

There are two normed subspaces of the normed space $\ell_\infty$ that are of importance in functional analysis. The space $c$ is the space of all convergent sequences, and $c_0$ is the space of all sequences converging to zero. Note that both spaces are endowed with the sup-norm inherited from the space $\ell_\infty$.

**Theorem 3.9.** *The spaces $c$ and $c_0$ are complete.*

*Proof.* We establish the claim of the theorem for the space $c_0$ (cf. Exercise 3.17). It suffices to prove that $c_0$ is a closed subset of $\ell_\infty$ (cf. Theorem 2.8).

Let $x^{(n)}$ be a sequence of vectors in $c_0$ that converges to $x \in \ell_\infty$. We need to show that $x \in c_0$. For $\varepsilon > 0$, there exists $N \in \mathbf{N}$ such that

$$\|x^{(n)} - x\|_\infty = \sup_{i \in \mathbf{N}} |x_i^{(n)} - x_i| < \varepsilon/2, \qquad \text{for all } n > N.$$

Hence

$$|x_i^{(N+1)} - x_i| < \varepsilon/2, \qquad \text{for all } i \in \mathbf{N}.$$

Because the sequence $(x_k^{(N+1)})$ converges to 0 as $k \to \infty$, there exists $N_1 \in \mathbf{N}$ such that

$$\left|x_k^{(N+1)}\right| < \varepsilon/2, \qquad \text{for all } k > N_1.$$

Thus for all $k > N_1$, we have

$$|x_k| = \left|x_k^{(N+1)} + (x_k - x_k^{(N+1)})\right| \le \left|x_k^{(N+1)}\right| + \left|x_k - x_k^{(N+1)}\right| < \varepsilon.$$

It follows that $x_k \to 0$, so $x \in c_0$. $\qquad\qquad\qquad\qquad\qquad\qquad\qquad\qquad$ $\square$

Because $c$ and $c_0$ are complete, they are Banach spaces.

The proof of the next theorem is left to the reader (cf. Exercise 3.18 and the proof of Theorem 3.7).

**Theorem 3.10.** *The spaces $c$ and $c_0$ are separable.*

Let $c_{00}$ be the set of sequences $(x_n)$ such that $x_n = 0$ for all but finitely many $n \in \mathbf{N}$. (Elements of $c_{00}$ are called *final* or *finitely supported* sequences.) It is clear that $c_{00}$ is a normed subspace of $c_0$ endowed with the sup-norm.

Unlike the spaces $c$ and $c_0$, the space $c_{00}$ is not a Banach space. Indeed, the sequence of vectors

$$a_n = \left(1, \frac{1}{2}, \dots, \frac{1}{n}, 0, 0, \dots\right), \qquad n \in \mathbf{N},$$

in the space $c_{00}$ is Cauchy, because for $m < n$ and $\varepsilon > 0$,

$$\|a_n - a_m\|_\infty = \frac{1}{m} < \varepsilon, \qquad \text{for } m, n > 1/\varepsilon.$$

Suppose that $(a_n)$ converges to the vector $a \in c_{00}$. Inasmuch as $a$ belongs to $c_{00}$, there exists $N \in \mathbf{N}$ such that

$$a = (x_1, \dots, x_N, 0, 0, \dots),$$

so $x_n = 0$ for $n > N$. Hence for $n > N$, we have

$$\|a_n - a\|_\infty = \left\|\left(1 - x_1, \dots, \tfrac{1}{N} - x_N, \tfrac{1}{N+1}, \dots, \tfrac{1}{n}, 0, 0, \dots\right)\right\|_\infty \ge \tfrac{1}{N+1},$$

which contradicts our assumption that $a_n \to a$ in $c_{00}$.

## 3.5 The Sequence Space $s$

The space $s$ is the vector space of all sequences $(x_n)$ with terms in $\mathbf{F}$. There are different methods of making $s$ into a topological vector space. We begin with a rather general approach to these methods that is also useful in other applications.

Let $\{p_j\}_{j \in J}$ be a family of seminorms on a vector space $X$. Finite intersections of open balls $B_j(a, r) = \{x \in X : p_j(x - a) < r\}$ form the basis of a topology on $X$ (cf. Exercise 3.23). In what follows, we assume that seminorms in $\{p_j\}_{j \in J}$ separate points in $X$, that is, for every two distinct vectors $x, y \in X$, there exists $j \in J$ such that $p_j(x) \neq p_j(y)$. Then the space $X$ is Hausdorff (cf. Exercise 3.24). If $J$ is a countable set, then we call the space $X$ a *polynormed space*.

In a polynormed space $X$ endowed with seminorms $\{p_i\}_{i \in \mathbf{N}}$, we define

$$d(x, y) = \sum_{i=1}^{\infty} \frac{1}{2^i} \frac{p_i(x - y)}{1 + p_i(x - y)}. \tag{3.10}$$

It is easy to verify that $d$ is a metric (cf. inequality (1.14) in Section 1.2). An important property of the metric $d$ is its translation invariance (cf. the distance $d_p$ in Section 3.2):

$$d(x + z, y + z) = d(x, y), \qquad \text{for all } x, y, z \in X.$$

Below, we use this property without referencing it explicitly. We also use the monotonicity of the real function $t/(1 + t)$, $t \geq 0$ (cf. the proof of Theorem 1.11), in some proofs in this section.

**Theorem 3.11.** *The topology $\mathcal{T}_1$ on a polynormed space $X$ defined by the family of seminorms $\{p_i\}_{i \in \mathbf{N}}$ coincides with the metric topology $\mathcal{T}_2$ defined by the metric $d$ on $X$ (cf. (3.10)).*

*Proof.* We need to prove two claims: (1) for every open set $U \in \mathcal{T}_1$ and every $x \in U$, there is an open ball $B \in \mathcal{T}_2$ that contains $x$ and is contained in $U$, and (2) for every open ball $B \in \mathcal{T}_2$ and every point $x$ in $B$, there is an open set $U \in \mathcal{T}_1$ such that $U \subseteq B$ and $x \in U$.

(1) It suffices to show that for every ball

$$B_k(a, R) = \{x \in X : p_k(x - a) < R\}$$

in $\mathcal{T}_1$ and point $x$ in this ball, there is a ball $B(x, r) \in \mathcal{T}_2$ that is contained in $B_k(a, R)$. Let $\delta_k = R - p_k(x - a) > 0$. Then for

$$r = \frac{1}{2^k} \frac{\delta_k}{1 + \delta_k}$$

and every $y \in B(x, r)$, we have

$$d(y, x) = \sum_{i=1}^{\infty} \frac{1}{2^i} \frac{p_i(y - x)}{1 + p_i(y - x)} < r = \frac{1}{2^k} \frac{\delta_k}{1 + \delta_k},$$

so

$$\frac{1}{2^k} \frac{p_k(y - x)}{1 + p_k(y - x)} < \frac{1}{2^k} \frac{\delta_k}{1 + \delta_k},$$

which is equivalent to

$$p_k(y - x) < \delta_k.$$

Then

$$p_k(y - a) \le p_k(y - x) + p_k(x - a) < \delta_k + p_k(x - a) = R,$$

that is, $y \in B_k(a, R)$. Hence $gB(x, r) \subseteq B_k(a, R)$.

(2) For $x \in B(a, R)$, we define $\delta = R - d(x, a) > 0$ and choose $n$ such that $n > \max\{1, \delta\}$ and

$$\sum_{i=n+1}^{\infty} \frac{1}{2^i} = \frac{1}{2^n} < \frac{\delta}{2}.$$

Let $y \in U = \bigcap_{i=1}^{n} B_i(x, \delta/(n - \delta))$, which is an open set in the topology $\mathcal{T}_1$. Then for every $1 \le i \le n$,

$$p_i(y - x) < \frac{\delta}{n - \delta}.$$

We have

$$d(y, x) = \sum_{i=1}^{\infty} \frac{1}{2^i} \frac{p_i(y - x)}{1 + p_i(y - x)} \le \sum_{i=1}^{n} \frac{1}{2^i} \frac{p_i(y - x)}{1 + p_i(y - x)} + \sum_{i=n+1}^{\infty} \frac{1}{2^i}$$

$$< \sum_{i=1}^{n} \frac{1}{2^i} \frac{\frac{\delta}{n-\delta}}{1 + \frac{\delta}{n-\delta}} + \sum_{i=n+1}^{\infty} \frac{1}{2^i} < \sum_{i=1}^{n} \frac{1}{2^i} \frac{\delta}{n} + \frac{\delta}{2} < \frac{\delta}{2} + \frac{\delta}{2} = \delta.$$

It follows that

$$d(y, a) \le d(y, x) + d(x, a) < \delta + d(x, a) = R,$$

so $y \in B(x, R)$. Hence $U = \bigcap_{i=1}^{n} B_i(x, \delta/(n - \delta)) \subseteq B(x, R)$ and $x \in U$. $\square$

**Theorem 3.12.** *In a polynormed space $X$, a sequence $(x_n)$ converges to $x$, that is, $d(x_n, x) \to 0$, if and only if $p_k(x_n - x) \to 0$ for every $k \in \mathbf{N}$.*

*Proof.* (Necessity.) Suppose that $x_n \to x$. Because

$$\frac{1}{2^k} \frac{p_k(x_n - x)}{1 + p_k(x_n - x)} \le d(x_n - x), \qquad k \in \mathbf{N},$$

we have $p_k(x_n - x) \to 0$ for every $k \in \mathbf{N}$. Indeed, because $d(x_n - x) \to 0$, for a given $\varepsilon > 0$ there exists $N$ such that

$$\frac{p_k(x_n - x)}{1 + p_k(x_n - x)} \le 2^k d(x_n - x) < \frac{\varepsilon}{1 + \varepsilon}, \qquad \text{for all } n > N.$$

Equivalently, $p_k(x_n - x) < \varepsilon$ for $n > N$. Hence $p_k(x_n - x) \to 0$ as $n \to \infty$, for every $k \in \mathbf{N}$.

(Sufficiency.) Let $p_k(x_n - x) \to 0$, for every $k \in \mathbf{N}$. For $\varepsilon > 0$, choose $M \in \mathbf{N}$ such that

$$\frac{1}{2^M} < \frac{\varepsilon}{2}$$

and $N \in \mathbf{N}$ such that

$$p_k(x_n - x) < \frac{\varepsilon}{2M}, \qquad \text{for } 1 \le k \le M \text{ and, } n > N.$$

Then

$$d(x_n, x) = \sum_{i=1}^{M} \frac{1}{2^i} \frac{p_i(x_n - x)}{1 + p_i(x_n - x)} + \sum_{i=M+1}^{\infty} \frac{1}{2^i} \frac{p_i(x_n - x)}{1 + p_i(x_n - x)}$$

$$\le \sum_{i=1}^{M} p_i(x_n - x) + \sum_{i=M+1}^{\infty} \frac{1}{2^i} < M \cdot \frac{\varepsilon}{2M} + \frac{1}{2^M} < \frac{\varepsilon}{2} + \frac{\varepsilon}{2} = \varepsilon,$$

for all $n > N$. Hence $x_n \to x$.  □

**Theorem 3.13.** *The polynormed space $X$ is a metric vector space.*

*Proof.* We need to show that (1) the operation of vector addition is continuous, and (2) the operation of multiplication by scalars is continuous.

(1) Let $x_n \to x$, $y_n \to y$. For a given $\varepsilon > 0$, choose $N \in \mathbf{N}$ such that

$$d(x_n, x) < \frac{\varepsilon}{2} \quad \text{and} \quad d(y_n, y) < \frac{\varepsilon}{2}, \qquad \text{for all } n > N.$$

Then

$$d(x_n + y_n, x + y) = d(x_n - x, y - y_n) \le d(x_n - x, 0) + d(0, y - y_n)$$
$$= d(x_n, x) + d(y_n, y) < \frac{\varepsilon}{2} + \frac{\varepsilon}{2} = \varepsilon,$$

for all $n > N$. It follows that $(x_n + y_n) \to (x + y)$, as required.

(2) Let $\lambda_n \to \lambda$ in $\mathbf{F}$ and $x_n \to x$ in $X$. By Theorem 3.12, $p_k(x_n - x) \to 0$ for every $k \in \mathbf{N}$. Then

$$p_k(\lambda_n x_n - \lambda x) = p_k(\lambda_n x_n - \lambda_n x + \lambda_n x - \lambda x)$$
$$\le |\lambda_n| p_k(x_n - x) + |\lambda_n - \lambda| p_k(x), \qquad \text{for all } k \in \mathbf{N}.$$

Because $\lambda_n \to \lambda$ and $p_k(x_n - x) \to 0$, the sequence on the right-hand side tends to zero as $n \to \infty$. Hence for every $k$, $p_k(\lambda_n x_n - \lambda x) \to 0$ as $n \to \infty$. By Theorem 3.12, $\lambda_n x_n \to \lambda x$.  □

We apply the above results to the vector space $s$ of all sequences $(x_k)$ with terms in the field $\mathbf{F}$. For

$$x = (x_1, x_2, \ldots, x_i, \ldots) \in s,$$

we define $p_i(x) = |x_i|$ for every $i \in \mathbf{N}$. Clearly, each $p_i$ is a seminorm on $s$. It is also clear that the family $\{p_i\}_{i \in \mathbf{N}}$ separates points in $s$. By Theorem 3.11, the metric

$$d(x, y) = \sum_{i=1}^{\infty} \frac{1}{2^i} \frac{|x_i - y_i|}{1 + |x_i - y_i|}$$

defines the same topology as the one defined by the family of seminorms $\{p_i\}_{i \in \mathbf{N}}$. By Theorem 3.12, convergence in the space $(s, d)$ is coordinatewise, that is, $x^{(n)} \to x$ in this space if and only if $x_i^{(n)} \to x_i$ for every $i \in \mathbf{N}$, and by Theorem 3.13, $(s, d)$ is a metric vector space.

Now we discuss some features of the space $s$ that are specific to this space. First, we show that $(s, d)$ is a complete metric space.

**Theorem 3.14.** *The space $(s, d)$ is a complete metric space.*

*Proof.* Let $(x^{(n)})$ be a Cauchy sequence in $s$ and let $\varepsilon > 0$. Fix $k \in \mathbf{N}$. There exists $N \in \mathbf{N}$ such that

$$d(x^{(m)}, x^{(n)}) < \frac{\varepsilon}{2^k(1 + \varepsilon)}, \qquad \text{for all } m, n > N.$$

It follows that

$$\frac{1}{2^k} \frac{|x_k^{(m)} - x_k^{(n)}|}{1 + |x_k^{(m)} - x_k^{(n)}|} < \frac{\varepsilon}{2^k(1 + \varepsilon)}, \qquad \text{for all } m, n > N,$$

which is equivalent to

$$|x_k^{(m)} - x_k^{(n)}| < \varepsilon, \qquad \text{for all } m, n > N.$$

By completeness of $\mathbf{F}$, the sequence $(x_k^{(n)})$ converges for every $k \in \mathbf{N}$. It follows from Theorem 3.12 that the sequence $(x^{(n)})$ converges in $s$. Hence $s$ is a complete space. $\qquad \square$

The proof of the next theorem employs a rather standard argument (cf. the proof of Theorem 3.7).

**Theorem 3.15.** *The space $s$ is separable.*

*Proof.* Let $M$ be the set of all sequences with rational coefficients that are terminated with zeros (so they belong to the space $c_{00}$). As the union of countably many countable sets, this set is a countable subset of $s$. We show that $\overline{M} = s$.

For a given $\varepsilon > 0$, we first choose $n$ such that $n > \max\{1, \varepsilon\}$ and

$$\sum_{i=n+1}^{\infty} \frac{1}{2^i} < \frac{\varepsilon}{2}.$$

Let $x = (x_n) \in s$. Now we choose rational numbers $r_1, \ldots, r_n$ such that

$$|x_i - r_i| < \frac{\varepsilon}{n - \varepsilon}, \qquad \text{for } 1 \leq i \leq n.$$

Then

$$\sum_{i=1}^{\infty} \frac{1}{2^i} \frac{|x_i - r_i|}{1 + |x_i - r_i|} = \sum_{i=1}^{n} \frac{1}{2^i} \frac{|x_i - r_i|}{1 + |x_i - r_i|} + \sum_{i=n+1}^{\infty} \frac{1}{2^i} \frac{|x_i - r_i|}{1 + |x_i - r_i|}$$

$$< \sum_{i=1}^{n} \frac{1}{2^i} \frac{\varepsilon}{n} + \sum_{n+1}^{\infty} \frac{1}{2^i} < \frac{\varepsilon}{2} + \frac{\varepsilon}{2} = \varepsilon.$$

Hence $M$ is dense in $s$. $\qquad\qquad\qquad\qquad\qquad\qquad\qquad\qquad\qquad\qquad\qquad$ □

A topological vector space is said to be *normable* if its topology is equivalent to the topology defined by a norm.

**Theorem 3.16.** *The metric vector space $s$ is not normable.*

*Proof.* Suppose that $\| \cdot \|$ is a norm on $s$ such that $\mathcal{T} = \mathcal{T}'$, where $\mathcal{T}$ is the norm topology and $\mathcal{T}'$ is the metric topology on $s$.

Let $B(0, 1) = \{x \in s : \|x\| < 1\}$ be the unit open ball in $\mathcal{T}$. Because $\mathcal{T} = \mathcal{T}'$, there is an open ball $B'(0, r) = \{x \in s : d(x, 0) < r\}$ in $\mathcal{T}'$ that is contained in $B(0, 1)$. Choose $n$ such that $1/2^n < r$ and let $e_n$ be a vector in $s$ with zeros everywhere but in the $n$th coordinate, where it is 1. We have

$$d(\lambda e_n, 0) = \frac{1}{2^n} \frac{|\lambda|}{1 + |\lambda|} \leq \frac{1}{2^n} < r, \qquad \text{for all } \lambda \in \mathbf{F}.$$

Hence $\lambda e_n \in B'(0, r) \subseteq B(0, 1)$ for all $\lambda \in \mathbf{F}$. This is a contradiction, because $\|\lambda e_n\| > 1$ for $|\lambda| > 1/\|e_n\|$. $\qquad\qquad\qquad\qquad\qquad\qquad\qquad\qquad$ □

For an alternative proof of this theorem, see Exercise 3.26.

## 3.6 Spaces of Continuous Functions

Let $C[0, 1]$ be the vector space of continuous real or complex functions on the unit interval $[0, 1]$. This vector space is a normed space with the norm defined by

$$\|x\| = \max_{t \in [0,1]} |x(t)|. \tag{3.11}$$

It is clear that $\|x\| = 0$ if and only is $x = 0$, and $\|\lambda x\| = |\lambda| \|x\|$ for all $\lambda \in \mathbf{F}$ and $x \in C[0, 1]$. The triangle inequality

$$\|x + y\| \leq \|x\| + \|y\|$$

follows immediately from the triangle inequality

$$|x(t) + y(t)| \le |x(t)| + |y(t)|, \qquad \text{for } t \in [0,1],$$

for real and complex numbers (cf. Exercise 3.11).

It is customary to denote the normed space $(C[0,1], \|\cdot\|)$ simply by $C[0,1]$, although we will consider other norms on the vector space $C[0,1]$. The norm defined by (3.11) is known as the *supremum norm* and sometimes denoted by $\|\cdot\|_\infty$ as in the case of the normed space $\ell_\infty$ (cf. (3.5)).

We begin by proving that $C[0,1]$ is a complete and separable normed space.

**Theorem 3.17.** *The normed space $C[0,1]$ is complete.*

*Proof.* Let $(x_n)$ be a Cauchy sequence of functions in $C[0,1]$. Then for a given $\varepsilon > 0$, there exists $N \in \mathbf{N}$ such that

$$\sup_{t \in [0,1]} |x_m(t) - x_n(t)| < \frac{1}{2}\varepsilon, \qquad \text{for all } m, n > N.$$

Hence for every $t \in [0,1]$,

$$|x_m(t) - x_n(t)| < \frac{1}{2}\varepsilon, \qquad \text{for all } m, n > N. \tag{3.12}$$

It follows that for every $t \in [0,1]$, the sequence of numbers $(x_n(t))$ is Cauchy and therefore converges to some number, which we denote by $x(t)$:

$$x_n(t) \to x(t), \qquad \text{for every } t \in [0,1].$$

By taking $n \to \infty$ in (3.12), we obtain

$$|x_m(t) - x(t)| \le \frac{1}{2}\varepsilon < \varepsilon, \qquad \text{for all } m > N \text{ and } t \in [0,1].$$

This shows that the sequence of continuous functions $(x_m)$ converges uniformly to the function $x$ on $[0,1]$. By a well-known theorem in analysis, $x$ is a continuous function and therefore belongs to $C[0,1]$. $\qquad\square$

To establish the separability of $C[0,1]$ (Theorem 3.18 below), we first prove the following lemma.

**Lemma 3.1.** *Let $\varepsilon > 0$ and suppose that $f$ is a continuous function on a closed interval $[a,b]$ such that*

$$|f(x) - f(y)| < \frac{\varepsilon}{5}, \qquad \text{for all } x, y \in [a,b].$$

*Let $h$ be a linear function on $[a,b]$ such that*

$$|h(a) - f(a)| < \frac{\varepsilon}{5} \quad and \quad |h(b) - f(b)| < \frac{\varepsilon}{5}.$$

*Then*

$$|f(x) - h(x)| < \varepsilon, \qquad for\ all\ x \in [a, b].$$

*Proof.* We have

$$|h(a) - h(b)| = |h(a) - f(a) + f(a) - f(b) + f(b) - h(b)|$$
$$\leq |h(a) - f(a)| + |f(a) - f(b)| + |f(b) - h(b)| < \frac{3}{5}\varepsilon.$$

Because $h$ is a linear function, it follows that

$$|h(x) - h(y)| < \frac{3}{5}\varepsilon, \qquad for\ all\ x, y \in [a, b].$$

Finally, we have

$$|f(x) - h(x)| = |f(x) - f(a) + f(a) - h(a) + h(a) - h(x)|$$
$$\leq |f(x) - f(a)| + |f(a) - h(a)| + |h(a) - h(x)|$$
$$< \frac{1}{5}\varepsilon + \frac{1}{5}\varepsilon + \frac{3}{5}\varepsilon = \varepsilon, \qquad for\ all\ x \in [a, b],$$

as required.                                                                    □

**Theorem 3.18.** *The normed space $C[a, b]$ is separable.*

*Proof.* The set $M$ of all continuous piecewise linear functions on $[0, 1]]$ with nodes at rational points in the plane is countable. To prove that $M$ is dense in $C[0, 1]$, it suffices to show that for every $f \in C[0, 1]$ and $\varepsilon > 0$, there exists $h \in M$ such that

$$|f(x) - h(x)| < \varepsilon, \qquad for\ all\ x \in [0, 1].$$

Inasmuch as $f$ is a uniformly continuous function on $[0, 1]$, there is $n \in \mathbf{N}$ such that

$$|f(x) - f(y)| < \frac{1}{5}\varepsilon, \quad for\ all\ x, y \in [0, 1]\ such\ that\ |x - y| \leq \frac{1}{n}.$$

Let $\{0, 1/n, \ldots, k/n, \ldots, 1\}$ be a partition of $[0, 1]$ into $n$ equal subintervals. For each point $k/n$, choose a rational number $h(k/n)$ such that

$$|h(k/n) - f(k/n)| < \frac{1}{5}\varepsilon.$$

The points $(k/n, h(k/n))$, $k = 0, 1, \ldots, n$, are nodes of a continuous piecewise linear function $h$ on $[0, 1]$. By Lemma 3.1,

3.6 Spaces of Continuous Functions

$$|f(x) - h(x)| < \varepsilon, \quad \text{over each interval } [k/n, (k+1)/n].$$

The result follows.                                              □

The normed space $C[a, b]$ is defined as the vector space of continuous functions on the closed interval $[a, b]$ with the norm

$$\|x\| = \max_{t \in [a,b]} |x(t)|.$$

It can be shown that this space is isomorphic to $C[0, 1]$ (cf. Exercise 3.27).

In a more general setting, let $C(K)$ be the vector space of continuous functions on a compact topological space $K$. By Theorem 2.25, the function

$$\|x\| = \max_{t \in K} |x(t)|, \qquad x \in C(K),$$

is well defined. It is a norm on $C(K)$. The normed space $C(K)$ is a Banach space (cf. Exercise 3.29).

A real- or complex-valued function $f$ on the interval $[0, 1]$ is of the *differentiability class* $C^n$ if the derivative $f^{(n)}$ exists and is continuous on $[0, 1]$. The vector space of all functions of class $C^n$ is denoted by $C^n[0, 1]$. It is a normed space if the norm is defined by

$$\|x\| = \sum_{k=1}^{n} \sup_{t \in [0,1]} |x^{(k)}(t)|, \qquad x \in C^n[0, 1]$$

(cf. Exercises 3.28 and 3.29)).

Let $C(0, 1)$ be the vector space of continuous real- or complex-valued functions on the open interval $(0, 1)$. We cannot use the supremum norm (3.11) on this space, because there are unbounded continuous functions on $(0, 1)$. However, we can utilize the approach outlined at the beginning of Section 3.5 to introduce a polynormed structure on $C(0, 1)$. For $k \geq 3$, we define

$$p_k(x) = \sup_{t \in I_k} |x(t)|, \qquad \text{where } I_k = [1/k, (k-1)/k]. \qquad (3.13)$$

It is not difficult to show that each function $p_k$ is indeed a seminorm and that the family $\{p_k\}_{k \geq 3}$ separates points in $C(0, 1)$ (cf. Exercise 3.30). Hence $C(0, 1)$ endowed with the family of seminorms $\{p_k\}_{k \geq 3}$ is a polynormed space that is also a metric vector space with the metric $d$ defined by (3.10).

In a similar manner, the vector space $C(\mathbf{R})$ of all continuous functions on $\mathbf{R}$ is made into a polynormed space by the family of seminorms defined by

$$p_k(x) = \sup_{|t| \leq k} |x(t)|, \qquad k \in \mathbf{N}.$$

When we say that $C[0, 1]$, $C^n[0, 1]$, and $C(K)$ are Banach spaces, we assume that they are endowed with respective versions of the supremum norm. However, there are other norms on, say, the space $C[0,1]$ that are of importance in analysis.

In the rest of this section, we consider the vector space $C[0,1]$ and the function

$$\|x\| = \int_0^1 |x(t)|\, dt. \tag{3.14}$$

This function is a norm. Indeed, because $|x(t)|$ is a nonnegative continuous real-valued function on $[0,1]$, it follows that $\int_0^1 |x(t)|\, dt = 0$ if and only if $x$ is the zero function on $[0,1]$ (cf. Exercise 3.31). It is clear that $\|\lambda x\| = |\lambda|\|x\|$. The triangle inequality for $\|\cdot\|$ follows immediately from the triangle inequality for the modulus function on $\mathbf{F}$. Hence $(C[0,1], \|\cdot\|)$ is a normed space. However, it is not a Banach space, because it is not complete. To show this, it suffices to produce a Cauchy sequence in $(C[0,1], \|\cdot\|)$ that does not converge to a point in this space.

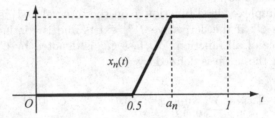

**Fig. 3.1** The function $x_n$.

For $n \geq 2$, we define a sequence $(x_n)$ of functions on the interval $[0,1]$ by

$$x_n(t) = \begin{cases} 0, & \text{if } 0 \leq t < 0.5, \\ n(t-0.5), & \text{if } 0.5 \leq t < a_n, \text{where } a_n = 0.5 + (1/n), \\ 1, & \text{if } a_n \leq t \leq 1 \end{cases}$$

(cf. Fig. 3.1). It is easy to verify that

$$\|x_m - x_n\| = \int_0^1 |x_m(t) - x_n(t)|\, dt = \left| \frac{1}{2m} - \frac{1}{2n} \right|,$$

so for $\varepsilon > 0$, we have

$$\|x_m - x_n\| < \varepsilon$$

if $m, n > 1/(2\varepsilon)$. Hence $(x_n)$ is a Cauchy sequence.

Suppose that the sequence $(x_n)$ converges to a continuous function $x$ in the normed space $C[0,1]$, that is, $\|x_n - x\| \to 0$. We have

$$\|x_n - x\| = \int_0^1 |x_n(t) - x(t)|\, d$$

$$= \int_0^{0.5} |x(t)|\, dt + \int_{0.5}^{a_n} |x_n(t) - x(t)|\, dt + \int_{0.5}^1 |x(t) - 1|\, dt - \int_{0.5}^{a_n} |x(t) - 1|\, dt.$$

Because $a_n \to 0.5$, the second and the last integrals on the right-hand side of the above equation converge to zero. Inasmuch as the left-hand side of the equation converges to zero, we have

$$\int_0^{0.5} |x(t)|\, dt = \int_{0.5}^1 |x(t) - 1|\, dt = 0.$$

Because $x$ is assumed to be a continuous function on $[0, 1]$, it follows that $x(t) = 0$ on $[0, 0.5]$ and $x(t) = 1$ on $[0.5, 1]$ (cf. Exercise 3.31), which is impossible. Hence the Cauchy sequence $(x_n)$ does not converge in the normed space $C[0, 1]$ endowed with the norm defined by (3.14), that is, this space is not a Banach space.

## 3.7 The $BV[a, b]$ Spaces

Let $x$ be a function on $[a, b]$. Observe that over a subinterval $[c, d]$ of $[a, b]$, the values of the function $x$ "change" or "vary" from $x(c)$ to $x(d)$, so the "change" in values is $|x(d) - x(c)|$. This observation motivates the following definition.

**Definition 3.5.** Let $x$ be a real- or complex-valued function on the interval $[a, b]$ and let $P = \{t_0, t_1, \ldots, t_n\}$ be a partition of this interval, that is,

$$a = t_0 < t_1 < \cdots < t_n = b.$$

The *variation of $x$ on $[a, b]$ with respect to $P$* is given by

$$V_a^b(x, P) = \sum_{k=1}^n |x(t_k) - x(t_{k-1})|.$$

If there is a constant $M$ such that $V_a^b(x, P) < M$ for every partition $P$ of $[a, b]$, then we say that $x$ is a *function of bounded variation*, a *BV-function* for short, on $[a, b]$ and write

$$V_a^b(x) = \sup\{V_a^b(x, P) : P \text{ is a partition of } [a, b]\}.$$

In this case, the quantity $V_a^b(x)$ is called the *total variation* of $x$ over the interval $[a, b]$. The set of all BV-functions on $[a, b]$ is denoted by $BV[a, b]$.

If $x$ and $y$ are BV-functions on $[a, b]$, then for every partition $P$ of the interval $[a, b]$ we have

$$V_a^b(x + y, P) \le V_a^b(x, P) + V_a^b(y, P) \le V_a^b(x) + V_a^b(y),$$

because

$$|(x(t_k) + y(t_k)) - (x(t_{k-1}) + y(t_{k-1}))| \le |x(t_k) - x(t_{k-1})| + |y(t_k) - y(t_{k-1})|,$$

for $P = \{t_0, t_1, \ldots, t_n\}$. Therefore, by taking the supremum over partitions $P$, we obtain

$$V_a^b(x + y) \le V_a^b(x) + V_a^b(y), \qquad \text{for } x, y \in BV[a, b].$$

Furthermore, for every function $x$ and every $\lambda \in \mathbf{F}$, we have

$$V_a^b(\lambda x, P) = |\lambda| V_a^b(x, P),$$

for every partition $P$ of $[a, b]$. If $x$ is a BV-function, then the above equation yields

$$V_a^b(\lambda x) = |\lambda| V_a^b(x), \qquad \text{for } \lambda \in \mathbf{F} \text{ and } x \in BV[a, b].$$

It follows that $BV[a, b]$ is a vector space over the field $\mathbf{F}$ and $V_a^b(x)$ is a seminorm on this space.

The following lemma shows that $V_a^b(x)$ is not a norm on the space $BV[a, b]$. The proof is left to the reader (cf. Exercise 3.32).

**Lemma 3.2.** $V_a^b(x) = 0$ *if and only if* $x$ *is a constant function on* $[a, b]$.

We use the result of Lemma 3.2 to prove the following theorem.

**Theorem 3.19.** *The function*

$$\|x\| = |x(a)| + V_a^b(x) \tag{3.15}$$

*is a norm on the vector space* $BV[a, b]$.

*Proof.* Note that $|x(a)|$ is a seminorm on $BV[a, b]$. Therefore, $\|x\|$ defined by (3.15) is also a seminorm on $BV[a, b]$ (cf. Exercise 3.4).

By Lemma 3.2, $\|x\| = 0$ if and only if $x$ is the zero function on $[a, b]$. Hence $\|\cdot\|$ is a norm on $BV[a, b]$. $\qquad\qquad\qquad\qquad\qquad\qquad\qquad\qquad\qquad\square$

We denote by the same symbol $BV[a, b]$ the normed space of functions of bounded variation on the interval $[a, b]$, where the norm is defined by (3.15), and show that this normed space is a Banach space.

First, we prove that a function of bounded variation is bounded by its norm.

**Lemma 3.3.** *Let $x$ be a BV-function on $[a,b]$. Then for every $t \in [a,b]$,*

$$|x(t)| \leq |x(a)| + V_a^b(x).$$

*Proof.* The inequality is trivial for $t = a$. For $t \in (a,b)$, let $P = \{a,t,b\}$, and for $t = b$, let $P = \{a,b\}$. Then

$$|x(t) - x(a)| \leq |x(t) - x(a)| + |x(b) - x(t)| = V_a^b(x,P) \leq V_a^b(x).$$

From this we deduce

$$|x(t)| = |x(a) + x(t) - x(a)| \leq |x(a)| + |x(t) - x(a)| \leq |x(a)| + V_a^b(x),$$

which is the desired result.                                                    □

**Theorem 3.20.** *The normed space $BV[a,b]$ is complete.*

*Proof.* Let $(x_n)$ be a Cauchy sequence in $BV[a,b]$. Then for $\varepsilon > 0$, there exists $N \in \mathbf{N}$ such that

$$\|x_m - x_n\| < \varepsilon/2, \qquad \text{for all } m, n > N.$$

By Lemma 3.3, for $t \in [a,b]$,

$$|x_m(t) - x_n(t)| \leq |x_m(a) - x_n(a)| + V_a^b(x_m - x_n) = \|x_m - x_n\| < \varepsilon/2,$$

for all $m, n > N$. Thus $(x_n(t))$ is a Cauchy sequence in $\mathbf{F}$ for every $t \in [a,b]$. Hence the sequence $(x_n)$ converges pointwise to some function $x$ on $[a,b]$:

$$x_n(t) \to x(t), \qquad t \in [a,b].$$

In what follows, we show that $x \in BV[a,b]$ and $\|x - x_n\| \to 0$.

For every partition $P$ of $[a,b]$, we have

$$\lim_{m \to \infty} \big[ |x_m(a) - x_n(a)| + V_a^b[(x_m - x_n, P)] \big]$$

$$= |x(a) - x_n(a)| + V_a^b(x - x_n), P),$$

because $x_m \to x$ pointwise. Inasmuch as

$$|x_m(a) - x_n(a)| + V_a^b[(x_m - x_n, P)] \leq \|x_m - x_n\| < \varepsilon/2,$$

we have

$$|x(a) - x_n(a)| + V_a^b(x - x_n), P) \leq \varepsilon/2 < \varepsilon \qquad \text{for all } n > N.$$

We conclude that $x - x_n$ is in the space $BV[a,b]$ for all $n > N$. Because $x_n \in BV[a,b]$, we have $x = (x - x_n) + x_n \in BV[a,b]$. From the last displayed inequality, we obtain

$$\|x - x_n\| = |x(a) - x_n(a)| + V_a^b(x - x_n) < \varepsilon \qquad \text{for all } n > N,$$

so $\|x - x_n\| \to 0$. Hence $BV[a, b]$ is a Banach space. $\qquad\square$

All spaces $BV[a, b]$ are inseparable. We prove this for the space $BV[0, 1]$.

**Theorem 3.21.** *The space $BV[0, 1]$ is inseparable.*

*Proof.* Let $\{x_\alpha\}_{\alpha \in (0,1)}$ be the family of functions on $[0, 1]$ defined by

$$x_\alpha(t) = \begin{cases} 0, & \text{if } t \in [0, \alpha), \\ 1, & \text{if } t \in [\alpha, 1]. \end{cases}$$

It is not difficult to verify that all functions $x_\alpha$ belong to the space $BV[0, 1]$ and

$$\|x_\alpha - x_\beta\| = 2, \qquad \text{if } \alpha \neq \beta.$$

The family $\{B(x_\alpha, 1)\}_{\alpha \in [0,1]}$ of disjoint open balls is uncountable. It follows that $BV[0, 1]$ cannot contain a dense countable set (cf. the proof of Theorem 3.7). $\qquad\square$

We conclude this section by introducing a subspace of the space $BV[a, b]$ that is of particular interest in functional analysis. Namely, we define

$$NBV[a, b] = \{x \in BV[a, b] : x(a) = 0\}.$$

It is clear that $NBV[a, b]$ is a normed subspace of $BV[a, b]$.

To show that $NBV[a, b]$ is a Banach space, it suffices to prove that it is closed in $BV[a, b]$. Let $(x_n)$ be a sequence in $NBV[a, b]$ converging to some $x \in BV[a, b]$, that is, $\|x_n - x\| \to 0$. Then

$$|x(a)| + V_a^b(x_n - x) \to 0,$$

implying that $x(a) = 0$. Hence $x \in NBV[a, b]$, which is the desired result.


## Notes


A topological vector space is not necessarily metrizable. An example is given by the vector space $X = \mathbf{R}^{\mathbf{R}}$ of all real-valued function on $\mathbf{R}$ endowed with the product topology, or equivalently, by the topology of pointwise convergence (Munkres 2000, p. 133).

By Theorems 3.17 and 3.18, $C[0, 1]$ is a separable Banach space. In some sense this space is universal in the class of separable Banach spaces. Namely, every (real) separable Banach space is isomorphic to a closed subspace of the (real) space $C[0,1]$ (Banach 1987, Ch. XI, §8).

The space $\ell_\infty$ can be considered the universal space in the class of separable metric spaces (cf. Exercise 3.22).

The theory of normed sequence spaces is rich with very deep results (Lindenstrauss and Tzafriri 1977). In this chapter and Section 5.1 we just touch the tip of the iceberg.

## Exercises

**3.1.** Let $X$ be a vector space and $d$ a metric on $X$. Show that a pair $(X,d)$ is a metric vector space if for sequences $(x_n)$ and $(y_n)$ of vectors in $X$ and sequences $(\lambda_n)$ of scalars in $\mathbf{F}$, these conditions hold:

(a)  $x_n \to x$, $y_n \to y$  imply  $(x_n + y_n) \to (x+y)$,
(b)  $\lambda_n \to \lambda$, $x_n \to x$  imply  $\lambda_n x_n \to \lambda x$.

**3.2.** For a norm $\|\cdot\|$ on a vector space $X$, show that

(a) $\|0\| = 0$,
(b) $\|x\| > 0$  if  $x \neq 0$

(cf. Exercise 2.7).

**3.3.** Let $d$ be the discrete metric on a nontrivial vector space $X$. Show that $(X,d)$ is not a metric vector space.

**3.4.** Show that the sum of two seminorms is a seminorm.

**3.5.** Let $B = \{x_i\}_{i \in J}$ be a Hamel basis in a vector space $X$. Show that

$$\|x\| = |\lambda_1| + \cdots + |\lambda_n|,$$

for $x = \lambda_1 x_{i_1} + \cdots + \lambda_n x_{i_n}$, is a norm on $X$.

**3.6.** Show that the seminorm $p$ on a seminormed space $(X,p)$ is a continuous function. Also show that the norm is a continuous function on a normed space.

**3.7.** Prove the triangle inequality for the normed space $\ell_\infty^n$.

**3.8.** Show that for $x = (x_1, \ldots, x_n) \in \mathbf{F}^n$,

$$\|x\|_\infty = \lim_{p \to \infty} \|x\|_p.$$

**3.9.** Complete the proof of Theorem 3.3.

**3.10.** (a) Show that two metric spaces $(X, d)$ and $(X, d')$ are homeomorphic if and only if every open ball in $(X, d)$ contains an open ball centered at the same point in $(X, d')$ and conversely.

(b) Show that two metric vector spaces $(X, d)$ and $(X, d')$ are isomorphic if and only if every open ball centered at the zero vector in the first space contains an open ball centered at the zero vector in the second space and conversely.

**3.11.** Let $A$ and $B$ be bounded sets of real numbers. Prove that

$$\sup(A + B) = \sup A + \sup B.$$

**3.12.** Prove that $\|\cdot\|_\infty$ defined by (3.5) is a norm (cf. Exercise 3.11).

**3.13.** Prove that the spaces $\ell_p^n$ for $p \in (0, \infty) \cup \{\infty\}$ and $\ell_p$ for $0 < p < 1$ are complete.

**3.14.** Prove that the spaces $\ell_p^n$ for $p \in (0, \infty) \cup \{\infty\}$ and $\ell_p$ for $0 < p < 1$ are separable.

**3.15.** Prove that for $p < q$,

$$\sum_k |x_k|^p < \infty \quad \text{implies} \quad \sum_k |x_k|^q < \infty.$$

Deduce the result of Theorem 3.8. Prove the second claim of Theorem 3.8.

**3.16.** Show that $\ell_p$ is not a normed subspace of $\ell_q$ for $p \neq q$.

**3.17.** Prove that the space $c$ is complete (cf. Theorem 3.9).

**3.18.** Prove Theorem 3.10 (cf. the proof of Theorem 3.7).

**3.19.** Let $B(S)$ be the vector space of all bounded $\mathbf{F}$-valued functions on a nonempty set $S$. Define

$$\|x\| = \sup\{x(t) : t \in S\}.$$

Prove that $(B(S), \|\cdot\|)$ is a complete normed space (cf. Theorem 3.5).

**3.20.** (a) Let $bv$ be the set of all sequences $x = (x_n)$ such that

$$\|x\|_{bv} = |x_1| + \sum_{k=1}^\infty |x_{k+1} - x_k| < \infty.$$

(A sequence in $bv$ is said to be of *bounded variation* (cf. Sect. 3.7).) Show that $bv$ is a normed vector space with the norm $\|\cdot\|_{bv}$.

(b) Let $bv_0$ be the set of vectors $x = (x_n) \in bv$ such that $\lim x_n = 0$. Show that $bv$ is a normed vector space with the norm defined by

$$\|x\|_{bv_0} = \sum_{k=1}^\infty |x_{k+1} - x_k|.$$

**3.21.** Show that for $1 \leq p < \infty$,

$$c_{00} \subseteq \ell_p \subseteq c_0 \subseteq c \subseteq \ell_\infty,$$

and all inclusions are proper.
Also show that

(a) $c_{00}$ is dense in the space $c_0$,
(b) $c_{00}$ is dense in $\ell_p$,
(c) $c_{00}$ is not dense in $c$,
(d) $c_{00}$ is not dense in $\ell_\infty$,

in the topology defined by the sup-norm, $\|\cdot\|_\infty$.

**3.22.** Prove that every separable metric space $M$ is isometric to a subspace of the space $\ell_\infty$.

**3.23.** Prove that finite intersections of open balls in a polynormed space $X$ form a basis for a topology on $X$. (Hint: Use Theorem 2.22.)

**3.24.** If a family $\{p_j\}_{j \in J}$ of seminorms on a vector space $X$ separates distinct points of $X$, show that the resulting topological vector space is Hausdorff.

**3.25.** If a metric vector space $(X, d)$ is normable, show that

(a) $d(x + z, y + z) = d(x, y)$,
(b) $d(\lambda x, \lambda y) = |\lambda|\, d(x, y)$,

for all $x, y, z \in X$ and $\lambda \in \mathbf{F}$. Deduce that the metric vector space $s$ is not normable (cf. Theorem 3.16).

**3.26.** Show that every open ball in the space $s$ centered at zero contains a subspace of $s$ that is isomorphic as a vector space to the space $s$ itself. (Hint: Modify the proof of Theorem 3.16.)

**3.27.** Show that the normed spaces $C[a, b]$ and $C[0, 1]$ are isomorphic.

**3.28.** Show that

$$\|x\| = \sum_{k=1}^{n} \sup_{t \in [0,1]} |x^{(k)}(t)|$$

is a norm on $C^n[0, 1]$.

**3.29.** Let $K$ be a compact topological space. Prove that the normed spaces $C(K)$ and $C^n[0, 1]$ are Banach spaces.

**3.30.** Show that the functions defined in (3.13) are seminorms on $C(0, 1)$ separating points in this space.

**3.31.** Prove that a nonnegative continuous real-valued function $f(x)$ on the interval $[a, b]$ is the zero function on $[a, b]$ if $\int_a^b f(x)\, dx = 0$.

**3.32.** Prove Lemma 3.2.

# Chapter 4
# Normed Spaces

The first two sections of this chapter are concerned with basic properties of normed spaces and linear operators on them. The concept of a Schauder basis is central in Section 4.1. These bases are useful tools in investigating sequence spaces and linear functionals on them. Section 4.2 is concerned mostly with bounded operators and the space of bounded operators from a normed space into another normed space. The central result of this section establishes the equivalence of continuity and boundedness of a linear operator. Although the concept of the dual space is introduced in Section 4.2, we postpone studies of linear functionals and duality until Chapter 5. Several examples of linear operators are also presented in Section 4.2.

Properties of finite-dimensional normed spaces are the subject of Section 4.3. The central result of this section is the isomorphism of all $n$-dimensional normed spaces. Finally, in Section 4.4, we prove Riesz's lemma and use it to characterize finite-dimensional normed spaces in terms of the compactness property of the unit ball. A geometric application of Riesz's lemma is found at the end of this section.

## 4.1 Properties of Normed Spaces

Let $X$ be a vector space with a norm $\| \cdot \|$ on it. Recall (cf. Section 3.1) that convergence in $(X, \| \cdot \|)$ is the convergence of sequences in the metric space $(X, d)$ with the distance function $d(x, y) = \|x - y\|$, $x, y \in X$. Namely, a sequence $(x_n)$ in a normed space $X$ converges to $x \in X$ if $\|x_n - x\| \to 0$ as $n \to \infty$. Then we write $x_n \to x$ and call $x$ the *limit* of $(x_n)$. A sequence $(x_n)$ in $X$ is *Cauchy* if for every $\varepsilon > 0$ there exists $N \in \mathbf{N}$ such that

$$\|x_m - x_n\| < \varepsilon, \qquad \text{for all } m, n > N$$

(cf. Section 2.7).

© Springer International Publishing AG, part of Springer Nature 2018
S. Ovchinnikov, *Functional Analysis*, Universitext,
https://doi.org/10.1007/978-3-319-91512-8_4

Inasmuch as a normed space $X$ is a vector space, we can introduce series in $X$ as follows.

Let $(x_k)$ be a sequence in $X$. We define a sequence $(s_n)$ of *partial sums*

$$s_n = x_1 + \cdots + x_n, \qquad \text{where } n \in \mathbf{N}.$$

If the sequence $(s_n)$ converges to some $s \in X$, then we say that the *series*

$$x_1 + \cdots + x_k + \cdots = \sum_{k=1}^{\infty} x_k$$

*converges* to $s$. In this case, we also say that the series $\sum_{k=1}^{\infty} x_k$ is *convergent*, and we call $s$ the *sum* of the series and write

$$s = \sum_{k=1}^{\infty} x_k = x_1 + \cdots + x_k + \cdots.$$

A series $\sum_{k=1}^{\infty} x_k$ in a normed space $X$ is said to be *absolutely convergent* if the series $\sum_{k=1}^{\infty} \|x_k\|$ converges in $\mathbf{R}$. A simple example from calculus of the series $\sum_{k=1}^{\infty} (-1)^{k+1}/k$ in the 1-dimensional normed space $X = \mathbf{R}^1$ shows that convergence does not imply absolute convergence. (The harmonic series $\sum_{k=1}^{\infty} 1/k$ diverges.) It is shown in calculus that absolute convergence of a series with real terms implies its convergence. However, that is not true in general normed spaces (cf. Exercises 4.1 and 4.2).

The term "basis" has a multitude of meanings in mathematics. The concept of convergence in a normed space $X$ is used to define a Schauder basis in this space.

**Definition 4.1.** A sequence of vectors $(e_n)$ in a normed space $X$ is said to be a *Schauder basis* (or *basis*) for $X$ if for every $x \in X$, there is a unique sequence of scalars $(\lambda_n)$ such that the series $\sum_{k=1}^{\infty} \lambda_k e_k$ converges to $x$, that is,

$$x = \sum_{k=1}^{\infty} \lambda_k e_k.$$

The series on the right-hand side of the above identity is called the *expansion* of $x$ with respect to the basis $(e_n)$.

Some sequence spaces, for instance $\ell_p$ for $p \in [1, \infty)$, have Schauder basis (cf. Exercise 4.3). However, as the following theorem suggests, the space $\ell_\infty$ does not have a basis (cf. Theorem 3.7).

**Theorem 4.1.** *If a normed space $X$ has a Schauder basis, then $X$ is separable.*

*Proof.* Let $(e_n)$ be a basis in the normed space $X$. For an $x \in X$, let $\sum_{k=1}^{\infty} \lambda_k e_k$ be its expansion. We define

$$M = \left\{ y = \sum_{k=1}^{n} \mu_k e_k : n \in \mathbf{N}, \ \mu_k \text{ is a rational number for all } k \in \mathbf{N} \right\}.$$

It is clear that $M$ is a countable set.

For $\varepsilon > 0$, there exists $N \in \mathbf{N}$ such that

$$\left\| x - \sum_{k=1}^{N} \lambda_k e_k \right\| < \frac{\varepsilon}{2}.$$

Because the rational numbers are dense in $\mathbf{F}$, there are rational numbers $\mu_k$, $1 \le k \le N$, such that

$$|\lambda_k - \mu_k| < \frac{\varepsilon}{2N \max\{\|e_k\| : 1 \le k \le N\}}.$$

Then for $y = \sum_{k=1}^{N} \mu_k e_k$, we have

$$\|x - y\| = \left\| x - \sum_{k=1}^{N} \lambda_k e_k + \sum_{k=1}^{N} (\lambda_k - \mu_k) e_k \right\|$$

$$\le \left\| x - \sum_{k=1}^{N} \lambda_k e_k \right\| + \sum_{k=1}^{N} |\lambda_k - \mu_k| \|e_k\| < \frac{\varepsilon}{2} + \sum_{k=1}^{N} \frac{1}{2N} \varepsilon = \varepsilon.$$

Because $y \in M$, it follows that $M$ is dense in $X$. $\qquad\square$

Unlike most special spaces introduced in Chapter 3, the normed space $c_{00}$ is not complete (cf. Section 3.4). However, this space is dense as a metric space in the complete metric space $c_0$ (cf. Exercise 3.21). Indeed, let $x = (x_1, \ldots, x_n, \ldots)$ be a vector in $c_0$, so $x_n \to 0$ as $n \to \infty$. Then for a given $\varepsilon > 0$, there exists $N \in \mathbf{N}$ such that $|x_n| < \varepsilon/2$ for all $n > N$. For $y = (x_1, \ldots, x_N, 0, 0, \ldots) \in c_{00}$,

$$\|x - y\|_\infty = \|(0, \ldots, 0, x_{N+1}, x_{N+2}, \ldots)\|_\infty = \sup_{n > N} |x_n| \le \frac{\varepsilon}{2} < \varepsilon.$$

Hence $c_{00}$ is dense in $c_0$, that is, the normed space $c_0$ is the metric completion of its normed subspace $c_{00}$. In this context, it is natural to say that $c_0$ is a Banach completion of the normed space $c_{00}$.

The case of the space $c_{00}$ is not unique. The next theorem shows that every normed space is isometric to a dense normed subspace of a Banach space.

**Theorem 4.2.** *Let $(X, \|\cdot\|)$ be a normed space. Then there exist a Banach space $\widetilde{X}$ and an isometry $T : X \to \widetilde{X}$ such that $T(X)$ is dense in $\widetilde{X}$. The space $\widetilde{X}$ is unique up to isometry of normed spaces.*

The space $\widetilde{X}$ in Theorem 4.2 is called the *Banach completion* of the normed space $X$. The reader is invited to supply details for the following proof. We use notation from Theorem 2.14.

*Proof.* By Theorem 2.14, there exist a complete metric space $(\widetilde{X}, \widetilde{d})$ and an isometry $T$ of the metric space $X$ into the metric space $\widetilde{X}$ such that $T(X)$ is dense in $\widetilde{X}$. To prove Theorem 4.2 it suffices to extend the vector addition and scalar multiplication operations from $X$ to $\widetilde{X}$ to make it a vector space, and then introduce a norm on $\widetilde{X}$ that makes it a Banach space.

The elements of the space $\widetilde{X}$ are equivalence classes of Cauchy sequences in $X$. Let $[(x_n)]$ and $[(b_n)]$ be points in $\widetilde{X}$. The sequence $(x_n + y_n)$ is Cauchy (why?) and therefore defines a point $[(x_n + y_n)]$ in $\widetilde{X}$. We define the sum of $[(x_n)]$ and $[(y_n)]$ in $\widetilde{X}$ by

$$[(x_n)] + [(y_n)] = [(x_n + y_n)].$$

To show that this sum is well defined, we need to show that if $(x_n) \sim (x_n')$ and $(y_n) \sim (y_n')$, then $(x_n + y_n) \sim (x_n' + y_n')$. This follows immediately from the inequality

$$\|(x_n + y_n) - (x_n' + y_n')\| \leq \|x_n - x_n'\| + \|y_n - y_n'\|.$$

Similarly, for $\lambda \in \mathbf{F}$, the product $\lambda[(x_n)]$ is defined as the equivalence class $[\lambda(x_n)]$. Again, this operation is well defined (does not depend on the choice of the representative of $[(x_n)]$).

It is not difficult to verify that the operations of addition and scalar multiplication defined above make $\widetilde{X}$ into a vector space. From these definitions, it also follows that on $T(X) \subseteq \widetilde{X}$ the operations of vector space induced from $\widetilde{X}$ coincide with those induced from $X$ by means of the isometry $T$.

We define $\|[(x_n)]\|_1 = \widetilde{d}(0, [(x_n)])$ for $[(x_n)] \in \widetilde{X}$. Using the isometry property of $T$, one can show that $\|\cdot\|_1$ is a norm on $\widetilde{X}$.

Finally, the uniqueness of the Banach completion $\widetilde{X}$ follows from the uniqueness of the metric completion of $X$. $\qquad\square$

At the end of Section 3.6, we proved that the vector space $C[0,1]$ endowed with the norm

$$\|x\| = \int_0^1 |x(t)|\, dt$$

(cf. 3.14) is not complete. The Banach completion of this normed space is the Banach space $L^1[0,1]$ of Lebesgue measurable functions $x$ on $[0,1]$ such that the Lebesgue integral of $|x|$ is finite.

Similarly, using Minkowski's inequality (1.17), one can show that the function

$$\|x\| = \left( \int_0^1 |x(t)|^p\, dt \right)^{1/p}$$

defines a norm on $C[0,1]$ for $p \geq 1$. The Banach completion of this (incomplete) normed space is the Banach space $L^p[0,1]$ of Lebesgue measurable functions $x$ on $[0,1]$ such that the Lebesgue integral of $|x|^p$ is finite.

The proofs of these assertions are beyond the scope of this book.

## 4.2 Linear Operators and Functionals

Let $X$ and $Y$ be normed spaces. It is customary to use the same notation $\|\cdot\|$ for norms in normed spaces unless it is necessary to distinguish between different norms on the same vector space. In functional analysis, linear maps $T : X \to Y$ are called *linear operators* (*linear functionals* if $Y = \mathbf{F}$). We adhere to the notation used in Section 2.4. Continuous linear operators are of special interest in functional analysis.

**Definition 4.2.** A linear operator $T$ from a normed space $X$ into a normed space $Y$ is said to be *continuous at the point* $x_0 \in X$ if for every $\varepsilon > 0$, there exists $\delta > 0$ such that

$$\|x - x_0\| < \delta \quad \text{implies} \quad \|Tx - Tx_0\| < \varepsilon, \qquad \text{for all } x \in X.$$

The operator $T : X \to Y$ is *continuous on* $X$ if it is continuous at every point in $X$.

Because normed spaces are metric vector spaces, this definition is a special case of Definition 2.9, and hence can be applied to arbitrary functions from the space $X$ into the space $Y$.

**Definition 4.3.** Let $X$ and $Y$ be normed spaces and $T : X \to Y$ a linear operator. The operator $T$ is said to be *bounded* if there exists $C > 0$ such that

$$\|Tx\| \leq C\|x\|, \qquad \text{for all } x \in X. \tag{4.1}$$

The following two theorems show that the concepts of continuity and boundedness are closely related.

**Theorem 4.3.** *If a linear operator $T : X \to Y$ is continuous at $x = 0$, then it is bounded.*

*Proof.* If $T$ is continuous at 0, then for $\varepsilon = 1$ there exists $\delta > 0$ such that

$$\|x\| < \delta \quad \text{implies} \quad \|Tx\| < 1, \qquad \text{for } x \in X.$$

Let $y$ be a nonzero vector in $X$. We have

$$\left\|\frac{\delta}{2\|y\|} y\right\| = \frac{\delta}{2} < \delta.$$

Then

$$\left\| T\left(\frac{\delta}{2\|y\|}\, y\right)\right\| = \frac{\delta}{2\|y\|}\|Ty\| < 1.$$

Hence

$$\|Ty\| \le \frac{2}{\delta}\,\|y\|, \qquad \text{for } y \in X, \text{ including } y = 0.$$

The result follows by setting $C = 2/\delta$ in Definition 4.3. $\qquad\qquad$ □

**Theorem 4.4.** *A bounded linear operator* $T : X \to Y$ *is uniformly continuous.*

*Proof.* If $T$ is bounded, then by Definition 4.3, there exists $C > 0$ such that

$$\|Tx - Ty\| = \|T(x - y)\| \le C\|x - y\|, \qquad \text{for all } x, y \in X.$$

For a given $\varepsilon > 0$, we set $\delta = \varepsilon/C$. Then for $x, y \in X$,

$$\|x - y\| < \delta \qquad \text{implies} \qquad \|Tx - Ty\| \le C\|x - y\| < \varepsilon.$$

It follows that $T$ is a uniformly continuous operator. $\qquad\qquad$ □

The results of Theorems 4.3 and 4.4 are often presented as the statement of the next theorem. The proof is left as an exercise (cf. Exercise 4.6).

**Theorem 4.5.** *A linear operator is bounded if and only if it is continuous.*

**Corollary 4.1.** *The null space* $\mathcal{N}(T)$ *of a bounded operator* $T : X \to Y$ *is closed.*

*Proof.* The set $\{0\}$ in $Y$ is closed. Inasmuch as $\mathcal{N}(T) = T^{-1}(\{0\})$ and $T$ is continuous, the result follows from Theorem 2.11 (cf. Exercise 4.4). $\qquad\qquad$ □

The converse of Corollary 4.1 does not hold in general, as the following example demonstrates.

*Example 4.1.* Let $T : c_{00} \to c_{00}$ be a linear operator defined by

$$T : (x_1, x_2, \ldots, x_n, \ldots) \mapsto (x_1, 2x_2, \ldots, nx_n, \ldots).$$

The null space of $T$ is the trivial subspace $\{0\}$, which is closed. For the vectors in the Schauder basis $(e_n)$ of $c_{00}$ (cf. Exercise 4.3), we have

$$\|e_n\| = 1 \qquad \text{and} \qquad \|Te_n\| = n, \qquad \text{for all } n \in \mathbf{N}.$$

Hence $T$ is an unbounded operator.

For linear functionals (so $Y = \mathbf{F}$) we have a stronger result than that of Corollary 4.1.

**Theorem 4.6.** *A linear functional on the normed space $X$ is bounded if and only if its null space is closed.*

*Proof.* (Necessity.) This is a special case of Corollary 4.1.

(Sufficiency.) Let $f$ be a linear functional on $X$ such that $\mathcal{N}(f)$ is a closed subspace of $X$. Suppose that $f$ is unbounded. Then for every $n \in \mathbf{N}$, there exists $x_n \in X$ such that $\|x_n\| = 1$ and $|f(x_n)| > n$. Note that $\mathcal{N}(f) \neq X$. Let $x \in X \setminus \mathcal{N}(f)$, so $f(x) \neq 0$, and define

$$y_n = x - \frac{f(x)}{f(x_n)} x_n \qquad n \in \mathbf{N}.$$

For every $n \in \mathbf{N}$ we have

$$f(y_n) = f(x) - \frac{f(x)}{f(x_n)} f(x_n) = 0.$$

Hence $y_n \in \mathcal{N}(f)$ for all $n \in \mathbf{N}$. Furthermore,

$$\|y_n - x\| = \left\| \frac{f(x)}{f(x_n)} x_n \right\| = \frac{|f(x)|}{|f(x_n)|} \|x\| < \frac{|f(x)|\|x\|}{n} \to 0,$$

as $n \to \infty$. Therefore, the sequence $(y_n)$ converges to $x$. Because $\mathcal{N}(f)$ is closed, we have $x \in \mathcal{N}(f)$, a contradiction. $\qquad\square$

The operator $T$ in Example 4.1 is unbounded. Here is another example of an unbounded operator.

*Example 4.2.* Let $X$ be the normed space of the polynomials on $[0,1]$ with the sup-norm. A *differentiation* operator $T$ is defined on $X$ by

$$Tx(t) = \frac{d}{dt} x(t), \qquad \text{for } x(t) \in X.$$

For $x_n(t) = t^n$, $n > 1$, we have $Tx_n(t) = nt^{n-1}$. It is clear that $\|x_n\| = 1$ and $\|Tx_n\| = n$. It follows that $T$ is an unbounded operator.

Unlike the differentiation operator, the integral operator in the next example is bounded.

*Example 4.3.* Here $X = C[0,1]$ is the normed space of continuous functions on $[0,1]$ with the sup-norm (cf. Section 3.6). For $x \in X$, the *integral* operator $T$ is defined by

$$(Tx)(t) = \int_0^1 k(t,u)x(u)\,du,$$

where $k$ is a continuous function on the closed square $[0,1] \times [0,1]$. The function $k$ is called the *kernel* of $T$. The operator $T$ is a linear operator mapping $C[0,1]$ to $C[0,1]$ (cf. Exercise 4.7).

Because $k$ is a continuous function on the closed square $[0,1]^2$, there is a constant $C > 0$ such that $|k(t,u)| \leq C$ for all $(t,u) \in [0,1]^2$. Since $|x(u)| \leq \|x\|$ for all $u \in [0,1]$, we have

$$\|Tx\| = \sup_{t \in [0,1]} \left| \int_0^1 k(t,u)x(u)\,du \right| \leq \sup_{t \in [0,1]} \int_0^1 |k(t,u)|\,|x(u)|\,du \leq C\|x\|.$$

Hence $T$ is bounded.

If $T : X \to Y$ is a bounded operator, then the set

$$\left\{ \frac{\|Tx\|}{\|x\|} : x \neq 0,\ x \in X \right\}$$

is bounded (cf. (4.1)). The supremum of this set is called the *operator norm* of $T$ and is denoted by $\|T\|$. Thus,

$$\|T\| = \sup_{\substack{x \in X \\ x \neq 0}} \frac{\|Tx\|}{\|x\|}.$$

It follows that

$$\|Tx\| \leq \|T\|\|x\|, \qquad \text{for all } x \in X.$$

This inequality will be used frequently without explicit reference to it.

Because

$$\frac{\|Tx\|}{\|x\|} = \left\| T\left( \frac{x}{\|x\|} \right) \right\| \quad \text{and} \quad \left\| \frac{x}{\|x\|} \right\| = 1, \qquad \text{for } x \neq 0,$$

we have an alternative definition of the operator norm:

$$\|T\| = \sup_{\substack{x \in X \\ \|x\|=1}} \|Tx\|.$$

The choice of the term "operator norm" can be justified as follows. Suppose that $X$ and $Y$ are normed spaces. We denote by $\mathcal{B}(X,Y)$ the set of all bounded operators from $X$ into $Y$. Clearly, it is a subset of the vector space $\mathcal{L}(X,Y)$ of all linear maps from $X$ to $Y$.

**Theorem 4.7.** *The set $\mathcal{B}(X,Y)$ is a vector subspace of $\mathcal{L}(X,Y)$. The function $T \mapsto \|T\|$ is a norm on $\mathcal{B}(X,Y)$.*

*Proof.* For $S, T \in \mathcal{B}(X,Y)$ we have

$$\|(S+T)(x)\| = \|Sx + Tx\| \leq \|Sx\| + \|Tx\| \leq \|S\|\|x\| + \|T\|\|x\|$$
$$= (\|S\| + \|T\|)\|x\|, \qquad \text{for all } x \in X.$$

It follows that $S + T$ belongs to $\mathcal{B}(X,Y)$. Moreover (cf. Exercise 4.8),

$$\|S+T\| = \sup_{\substack{x\in X \\ \|x\|=1}} \|(S+T)(x)\| = \sup_{\substack{x\in X \\ \|x\|=1}} \|Sx+Tx\|$$

$$\leq \sup_{\substack{x\in X \\ \|x\|=1}} (\|Sx\|+\|Tx\|) \leq \sup_{\substack{x\in X \\ \|x\|=1}} \|Sx\| + \sup_{\substack{x\in X \\ \|x\|=1}} \|Tx\| = \|S\|+\|T\|.$$

We verify now that for a bounded $T$ and $\lambda \in \mathbf{F}$, the operator $\lambda T$ is also bounded and $\|\lambda T\| = |\lambda|\|T\|$. Indeed,

$$\|(\lambda T)(x)\| = \|\lambda(Tx)\| = |\lambda|\|Tx\| \leq |\lambda|\|T\|\|x\|,$$

so $\lambda T$ is bounded. Furthermore,

$$\|\lambda T\| = \sup_{\substack{x\in X \\ \|x\|=1}} \|(\lambda T)(x)\| = |\lambda| \sup_{\substack{x\in X \\ \|x\|=1}} \|Tx\| = |\lambda|\|T\|.$$

Finally, if $\|T\| = 0$, then $Tx = 0$ for all $x \in X$. Hence $T = 0$. $\qquad\square$

In what follows, the space $\mathcal{B}(X,Y)$ is a normed space with the operator norm on it.

An important property of the normed space $\mathcal{B}(X,Y)$ is the result of the following theorem.

**Theorem 4.8.** *If $Y$ is a Banach space, then $\mathcal{B}(X,Y)$ is a Banach space.*

*Proof.* Let $(T_n)$ be a Cauchy sequence in $\mathcal{B}(X,Y)$. We need to show that this sequence converges to some operator $T \in \mathcal{B}(X,Y)$.

First, we define the limit $T$ (cf. the proof of Theorem 3.5). For every $x \in X$,

$$\|T_n x - T_m x\| = \|(T_n - T_m)x\| \leq \|T_n - T_m\|\|x\|.$$

Because $(T_n)$ is Cauchy, it follows that $(T_n x)$ is a Cauchy sequence in the Banach space $Y$. Hence it is convergent. We define

$$Tx = \lim T_n x, \qquad x \in X.$$

The operator $T$ is linear, because

$$T(\lambda u + \mu v) = \lim(T_n(\lambda u + \mu v)) = \lim(\lambda T_n u + \mu T_n v) = \lambda Tu + \mu Tv,$$

for all $\lambda, \mu \in \mathbf{F}$ and $u,v \in X$.

It remains to show that $T \in \mathcal{B}(X,Y)$ and $T_n \to T$.

By Theorem 2.7, the Cauchy sequence $(T_n)$ is bounded, so there exists $C > 0$ such that $\|T_n\| < C$ for all $n \in \mathbf{N}$. Therefore,

$$\|T_n x\| \leq \|T_n\|\|x\| \leq C\|x\|, \qquad \text{for all } x \in X.$$

By the continuity of the norm (cf. Exercise 3.6), we have

$$\|Tx\| = \lim \|T_n x\| \leq C\|x\|, \qquad \text{for all } x \in X.$$

Hence $T$ is bounded, so $T \in \mathcal{B}(X,Y)$.

Because $(T_n)$ is Cauchy, for a given $\varepsilon > 0$ there exists $N \in \mathbf{N}$ such that

$$\|T_n - T_m\| < \varepsilon/2, \qquad \text{for all } n > N.$$

Then for every $x \in X$, we have

$$\|T_n x - T_m x\| = \|(T_n - T_m)x\| \leq \|T_n - T_m\|\|x\| \leq \frac{\varepsilon}{2}\|x\|.$$

Again by the continuity of the norm, for every $x \in X$ and $n > N$, we have

$$\|T_n x - Tx\| = \lim_{m \to \infty} \|T_n x - T_m x\| \leq \frac{\varepsilon}{2}\|x\|,$$

that is,

$$\|T_n - T\| = \sup_{\substack{x \in X \\ x \neq 0}} \frac{\|T_n x - Tx\|}{\|x\|} \leq \frac{\varepsilon}{2} < \varepsilon.$$

Hence $T_n \to T$ in $\mathcal{B}(X,Y)$. $\qquad\qquad\qquad\qquad\qquad\qquad\qquad\qquad$ □

If $Y = \mathbf{F}$, then the space $\mathcal{B}(X, \mathbf{F})$ is a normed space of bounded linear functionals on the normed space $X$. It is denoted by $X^*$ and called the *dual space* of $X$. The norm of a functional $f : X \to \mathbf{F}$ is given by

$$\|f\| = \sup_{\substack{x \in X \\ x \neq 0}} \frac{|f(x)|}{\|x\|},$$

or equivalently,

$$\|f\| = \sup_{\substack{x \in X \\ \|x\|=1}} |f(x)|.$$

By Theorem 4.8, the dual space $X^*$ of a normed space $X$ is a Banach space.

We conclude this section with more examples of bounded linear operators.

*Example 4.4.* **The shift operators.** The mappings

$$T_l : (x_1, \ldots, x_n, \ldots) \mapsto (x_2, \ldots, x_{n+1}, \ldots)$$

and

$$T_r : (x_1, \ldots, x_n, \ldots) \mapsto (0, x_1, \ldots, x_{n-1}, \ldots)$$

on the space $\ell_p$, $p \geq 1$, are called the *left shift operator* and the *right shift operator*, respectively. These operators are bounded, and the norm of each operator is 1 (cf. Exercise 4.9).

*Example 4.5.* For $d = (d_n) \in \ell_\infty$, the *diagonal operator* $T_d : \ell_p \to \ell_p$ is defined by

$$T_d : (x_1, \ldots, x_n, \ldots) \mapsto (d_1 x_1, \ldots, d_n x_n, \ldots).$$

For $x = (x_n) \in \ell_p$, we have

$$\|T_d x\|_p = \left( \sum_{k=1}^{\infty} |d_k x_k|^p \right)^{1/p} \leq \sup_{k \in \mathbf{N}} |d_k| \left( \sum_{k=1}^{\infty} |x_k|^p \right)^{1/p} = \|d\|_\infty \|x\|_p.$$

Hence, $T_d$ is well defined and $\|T_d\| \leq \|d\|_\infty$. Let $(e_n)$ be the Schauder basis in $\ell_p$ (cf. Exercise 4.3). Clearly, $T_d e_n = d_n$ for $n \in \mathbf{N}$. We have

$$\|T_d\| = \sup_{\|x\|=1} \|T_d x\|_p \geq \sup_{n \in \mathbf{N}} \|T_d e_n\| = \|d\|_\infty.$$

It follows that $\|T_d\| = \|d\|_\infty$.

*Example 4.6.* On the space $C[0,1]$, the *operator of indefinite integration* is defined by

$$(Tx)(t) = \int_0^t x(u)\, du, \qquad 0 \leq t \leq 1,\ x \in C[0,1].$$

The integral on the right-hand side is a continuous function of the upper limit. Hence $T$ indeed maps the space $C[0,1]$ into itself. For $x \in C[0,1]$, we have

$$\|Tx\| = \sup_{t \in [0,1]} \left| \int_0^t x(u)\, du \right| \leq \sup_{t \in [0,1]} \int_0^t |x(u)|\, du \leq \int_0^1 |x(u)|\, du$$

$$\leq \sup_{u \in [0,1]} |x(u)| = \|x\|.$$

For the constant function $x = 1$ on $[0,1]$, we have

$$\|Tx\| = \sup_{t \in [0,1]} \left| \int_0^t 1\, du \right| = \sup_{t \in [0,1]} t = 1 = \|x\|.$$

It follows that $\|T\| = 1$.

*Example 4.7.* For a given function $k \in C[0,1]$, we define a *multiplication operator* $T_k$ on $C[0,1]$ as follows:

$$(T_k x)(t) = k(t) \cdot x(t), \qquad 0 \leq t \leq 1,\ x \in C[0,1]$$

(cf. Example 4.5). We have

$$\|T_k x\|_\infty = \sup_{t \in [0,1]} |k(t)||x(t)| \leq \sup_{t \in [0,1]} |k(t)| \cdot \sup_{t \in [0,1]} |x(t)| = \|k\|_\infty \|x\|_\infty.$$

On the other hand, for the constant function $x = 1$ on $[0,1]$, we have

$$\|T_k x\|_\infty = \sup_{t \in [0,1]} |k(t)| = \|k\|_\infty.$$

Hence $\|T_k\| = \|k\|_\infty$.

## 4.3 Finite-Dimensional Normed Spaces

We begin with a motivating example. Let $\|x\|$ and $\|x\|'$ be norms on the vector space $\mathbf{R}^2 = \{(x_1, x_2) : x_1 , x_2 \in \mathbf{R}\}$ defined by

$$\|x\| = \sqrt{x_1^2 + x_2^2} \qquad \text{and} \qquad \|x\|' = \sqrt{x_1^2/a^2 + x_2^2/b^2},$$

respectively, where $a > b > 0$. The first norm is the usual $\ell_2$-norm, and it is not difficult to verify that $\|\cdot\|'$ is indeed a norm on $\mathbf{R}^2$ (cf. Exercise 4.13). It is a simple algebraic exercise (cf. Exercise 4.14) to show that

$$\frac{1}{a}\|x\| \leq \|x\|' \leq \frac{1}{b}\|x\|, \qquad \text{for all } x \in \mathbf{R}^2.$$

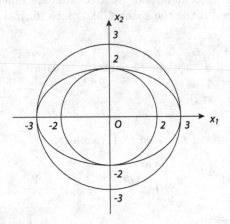

**Fig. 4.1** Three balls in $\mathbf{R}^2$.

The reader is invited to verify that the drawing in Figure 4.1 illustrates the relation between the norms $\|\cdot\|$ and $\|\cdot\|'$ for $a = 3$ and $b = 2$ by showing that

$$2B(0,1) \subseteq B'(0,1) \subseteq 3B(0,1),$$

where $B(0,1)$ and $B'(0,1)$ are the unit balls defined by $\|\cdot\|$ and $\|\cdot\|'$, respectively.

An important property of a finite-dimensional vector space $X$ is that similar inequalities can be established for every two norms on $X$, implying that all norms on $X$ define the same topology on $X$. In what follows, we investigate these and other properties of finite-dimensional normed spaces.

Let $\{x_1, \ldots, x_n\}$ be a linearly independent set of vectors in a normed space $X$. Obviously, the norm of a linear combination $x = c_1 x_1 + \cdots + c_n x_n$ is determined by the coefficients ("coordinates" of $x$) $c_1, \ldots, c_n$ of this linear combination. Although there is no "formula" expressing $\|c_1 x_1 + \cdots + c_n x_n\|$ in terms of $c_1, \ldots, c_n$, one can use the triangle inequality to establish a simple relation between these quantities.

Let $d = \max_k \|x_k\|$. Clearly, $d > 0$ and

$$\|c_1 x_1 + \cdots + c_n x_n\| \leq \sum_{k=1}^{n} |c_k| \|x_k\| \leq d \sum_{k=1}^{n} |c_k|,$$

for $c_k \in \mathbf{F}$, $1 \leq k \leq n$. In words, there is no "large" vector that is a linear combination of linearly independent vectors with "small" coefficients.

A more difficult task is to show that we cannot find a linear combination with "large" coefficients that represents a "small" vector. For this we have the following result:

**Lemma 4.1.** *For a linearly independent set of vectors $\{x_1, \ldots, x_n\}$ in $X$ there exists a real number $c > 0$ such that*

$$c \sum_{k=1}^{n} |c_k| \leq \|c_1 x_1 + \cdots + c_n x_n\|,$$

*for all $(c_1, \ldots, c_n)$.*

To prove Lemma 4.1, we need the result of the following lemma.

**Lemma 4.2.** *Let $\{x_1, \ldots, x_n\}$ be a finite subset of $X$. Then the function*

$$f(c_1, \ldots, c_n) = \|c_1 x_1 + \cdots + c_n x_n\|$$

*is uniformly continuous on $\ell_2^n$.*

*Proof.* We have (cf. Exercise 1 6)

$$|f(c_1,\ldots,c_n) - f(c_1',\ldots,c_n')| = \left| \|\sum_{k=1}^{n} c_k x_k\| - \|\sum_{k=1}^{n} c_k' x_k\| \right|$$

$$\leq \|\sum_{k=1}^{n} c_k x_k - \sum_{k=1}^{n} c_k' x_k\| = \|\sum_{k=1}^{n} (c_k - c_k') x_k\|$$

$$\leq \sum_{k=1}^{n} |c_k - c_k'| \|x_k\| \leq \max_{1 \leq k \leq n} \{\|x_k\|\} \sum_{k=1}^{n} |c_k - c_k'|$$

$$\leq \max_{1 \leq k \leq n} \{\|x_k\|\} \sqrt{n} \sqrt{\sum_{k=1}^{n} |c_k - c_k'|^2}$$

$$= \max_{1 \leq k \leq n} \{\|x_k\|\} \sqrt{n} \, \|(c_1,\ldots,c_n) - (c_1',\ldots,c_n')\|_2 \,.$$

We may assume that $\max_k \{\|x_k\|\} \neq 0$. For $\varepsilon > 0$, if

$$\|(c_1,\ldots,c_n) - (c_1',\ldots,c_n')\|_2 \leq \frac{\varepsilon}{\max_{1 \leq k \leq n} \{\|x_k\|\} \sqrt{n}},$$

then

$$|f(c_1,\ldots,c_n) - f(c_1',\ldots,c_n')| < \varepsilon.$$

Hence $f$ is a uniformly continuous function on $\ell_2^n$.  $\square$

Now we proceed with the proof of Lemma 4.1.

*Proof.* Clearly, we may assume that $\sum_{k=1}^{n} |c_k| \neq 0$. Then it suffices to prove that there exists a real number $c$ such that

$$\|\sum_{k=1}^{n} c_k x_k\| \geq c > 0,$$

for all $(c_1,\ldots,c_n)$ such that $\sum_{k=1}^{n} |c_k| = 1$. The set

$$S = \{(c_1,\ldots,c_n) : \sum_{k=1}^{n} |c_k| = 1\}$$

is closed and bounded in $\ell_2^n$ (cf. Exercise 4.15). Hence by the Bolzano–Weierstrass theorem (cf. Example 2.17), it is compact. By Lemma 4.2, the function

$$f(c_1,\ldots,c_n) = \|\sum_{k=1}^{n} c_k x_k\|$$

is continuous on $\ell_2^n$. Therefore, (cf. Theorem 2.21), it attains its minimum on the compact set $S$, that is, there exists $(c_1',\ldots,c_n') \in S$ such that

$$f(c_1,\ldots,c_n) \geq f(c_1',\ldots,c_n') = c \geq 0,$$

for all $(c_1, \ldots, c_n) \in S$. Suppose that $c = 0$. Then $\sum_{k=1}^{n} c_k' x_k = 0$, which implies that all the $c_k'$ are zero, because the set $\{x_1, \ldots, x_n\}$ is linearly independent. This contradicts the fact that $(c_1', \ldots, c_n') \in S$. Hence $c > 0$, which is the desired result. $\qquad\square$

The inequality in Lemma 4.2 is instrumental in establishing the result of Theorem 4.9. First, we define a concept motivated by an opening example in this section.

**Definition 4.4.** Two norms $\| \cdot \|$ and $\| \cdot \|'$ on a vector space $X$ are said to be *equivalent* if there are positive real numbers $a$ and $b$ such that

$$b\|x\| \leq \|x\|' \leq a\|x\|,$$

for all $x \in X$.

**Theorem 4.9.** *Let $X$ be a finite-dimensional vector space. Then every two norms $\| \cdot \|$ and $\| \cdot \|'$ on $X$ are equivalent.*

*Proof.* Let $\{e_1, \ldots, e_n\}$ be a basis of $X$, and $c > 0$ the number from Lemma 4.1 for the norm $\| \cdot \|$. For $x = \sum_{k=1}^{n} c_k e_k$, we have

$$\|x\|' = \Big\| \sum_{k=1}^{n} c_k e_k \Big\|' \leq \max_{1 \leq k \leq n} \{\|e_k\|'\} \sum_{k=1}^{n} |c_k|$$

$$\leq \max_{1 \leq k \leq n} \{\|e_k\|'\} \frac{1}{c} \Big\| \sum_{k=1}^{n} c_k e_k \Big\| = \max_{1 \leq k \leq n} \{\|e_k\|'\} \frac{1}{c} \|x\|.$$

Hence for $a = \max_k \{\|x_k\|\}/c > 0$, we have $\|x\|' \leq a\|x\|$. By symmetry, there exists a positive $b$ such that $\|x\| \leq (1/b)\|x\|'$. $\qquad\square$

**Corollary 4.2.** *Every two norms on a finite-dimensional vector space $X$ define the same topology on $X$.*

*Proof.* It can be easily seen that an open set in a metric space is a union of open balls centered at the points of the set. Hence two topologies on $X$ are equal if (and only if) every open ball in the first topology contains an open ball of the second topology, and conversely. The claim of the corollary follows immediately from this observation and Theorem 4.9. $\qquad\square$

The assumption that $X$ is a finite-dimensional vector space in Theorem 4.9 is essential, as the following example demonstrates.

*Example 4.8.* Let $X = c_{00}$ be the vector space of real sequences $x = (x_n)$ terminating with zeros, and let $\| \cdot \|_1$ and $\| \cdot \|_\infty$ be norms defined by

$$\|x\|_1 = \sum_{k=1}^{n} |x_k| \qquad \text{and} \qquad \|x\|_\infty = \sup_k |x_k|,$$

respectively. Suppose that there exists $a > 0$ such that $\|x\|_1 \leq a\|x\|_\infty$ for all $x \in X$. For $N \in \mathbf{N}$, let $x_N = (x_1, x_2, \ldots)$ be the vector defined by $x_k = 1$ for $1 \leq k \leq N$, and $x_k = 0$ otherwise. We have

$$N = \|x_N\|_1 \leq a\|x_N\|_\infty = a, \qquad \text{for all } N \in \mathbf{N},$$

which is clearly impossible. It follows that the norms $\|\cdot\|_1$ and $\|\cdot\|_\infty$ are not equivalent on the space $X$.

The result of the following theorem is also based on Lemma 4.1.

**Theorem 4.10.** *If a normed space $X$ is finite-dimensional, then every linear operator $T : X \to Y$ is bounded.*

*Proof.* Let $\{e_1, \ldots, e_n\}$ be a basis for the $n$-dimensional normed vector space $X$, and let $T$ be a linear operator from $X$ to some normed space $Y$. By the linearity of $T$ and Lemma 4.1, for every $x = \sum_{k=1}^{n} c_k e_k$ in $X$, we have

$$\|Tx\| = \left\| \sum_{k=1}^{n} c_k Te_k \right\| \leq \sum_{k=1}^{n} |c_k| \|Te_k\| \leq \max_{1 \leq k \leq n} \|Te_k\| \sum_{k=1}^{n} |c_k|$$

$$\leq \max_{1 \leq k \leq n} \|Te_k\| \frac{1}{c} \left\| \sum_{k=1}^{n} c_k e_k \right\| = \max_{1 \leq k \leq n} \|Te_k\| \frac{1}{c} \|x\|,$$

where $c > 0$ is the number from Lemma 4.1 for the norm $\|\cdot\|$. It follows that $T$ is bounded. $\qquad \qquad \square$

Recall that two topological vector spaces are isomorphic if they are isomorphic as vector spaces and homeomorphic as topological spaces under a linear map (cf. Definition 3.4). The relation "$X$ is isomorphic to $Y$" is an equivalence relation between normed spaces. If $T$ is an isomorphism of normed spaces $X$ and $Y$, then both $T$ and $T^{-1}$ are bounded operators (cf. Exercise 4.16).

Isomorphic normed spaces enjoy the following property.

**Theorem 4.11.** *If $X$ and $Y$ are isomorphic normed spaces, then $X$ is complete if and only if $Y$ is complete.*

*Proof.* Suppose that $X$ is a complete normed space and the space $Y$ is isomorphic to $X$. Let $T : X \to Y$ be an isomorphism and $(y_n)$ a Cauchy sequence in $Y$. Because $T^{-1}$ is bounded, we have

$$\|T^{-1}y_m - T^{-1}y_n\| = \|T^{-1}(y_m - y_n)\| \leq \|T^{-1}\| \cdot \|y_m - y_n\|.$$

Hence, $(T^{-1}y_n)$ is a Cauchy sequence in $X$. Inasmuch as $X$ is complete, $T^{-1}y_n \to T^{-1}y$ for some $y \in Y$. Then

$$y_n = T(T^{-1}y_n) \to T(T^{-1}y) = y,$$

because $T$ is a continuous mapping. Hence $Y$ is a complete space. The result follows by symmetry. □

As a consequence of Corollary 4.2, we obtain the following result.

**Theorem 4.12.** *Every two $n$-dimensional normed spaces are isomorphic.*

*Proof.* It suffices to show that an $n$-dimensional normed space $X$ with a norm $\|\cdot\|$ is isomorphic to the space $\ell_1^n$. Let $\{e_1, \ldots, e_n\}$ be a basis of $X$ and define

$$\|x\|' = \sum_{k=1}^{n} |c_k|$$

for $x = \sum_{k=1}^{n} c_k e_k \in X$. It can be readily verified that $\|\cdot\|'$ is a norm on $X$. By Corollary 4.2, the normed spaces $(X, \|\cdot\|)$ and $(X, \|\cdot\|')$ are isomorphic. Furthermore, the mapping $\ell_1^n \to X$ defined by

$$(c_1, \ldots, c_n) \mapsto c_1 e_1 + \cdots + c_n e_n$$

is an isomorphism of normed spaces $(\ell_1^n, \|\cdot\|_1)$ and $(X, \|\cdot\|')$. Indeed, it is clearly an isomorphism of vector spaces and also an isometry, and therefore a homeomorphism. □

**Theorem 4.13.** *Every finite-dimensional subspace $Y$ of a normed space $X$ is complete and therefore closed.*

*Proof.* It suffices to note that by Theorem 4.12, an $n$-dimensional subspace $Y$ is isomorphic to the complete space $\ell_2^n$ and then use Theorem 4.11. □

**Theorem 4.14.** *A subset of a finite-dimensional normed space $X$ is compact if and only if it is closed and bounded.*

*Proof.* By Theorem 2.18, it suffices to show that a closed and bounded subset $M$ of $X$ is compact. Let $T : X \to \ell_2^n$ ($n$ is the dimension of $X$) be an isomorphism (cf. Theorem 4.12). Then the set $T(M)$ is a closed and bounded subset of $\ell_2^n$, which is compact by the Bolzano–Weierstrass theorem. By Theorem 2.20, the set $M = T^{-1}(T(M))$ is compact. □

In conclusion of this section, we give an application of Theorem 4.13. First we prove a lemma.

**Lemma 4.3.** *Every proper finite-dimensional subspace $Y$ of a normed space $X$ is nowhere dense in the space $X$.*

*Proof.* By Theorem 4.13, $Y$ is closed. Suppose that an open ball $B(x,r)$ is contained in $Y$. Let $y$ be a vector in $X \setminus Y$ and let

$$z = x + (y - x)\frac{r}{2\|y - x\|}.$$

Because $Y$ is a subspace and $x \in Y$, it follows that $z \notin Y$. On the other hand,

$$\|z - x\| = r/2 < r,$$

so $z \in B(x,r) \subseteq Y$, whence the required result. $\qquad\square$

Note that the assumption that $Y$ is finite-dimensional is essential in the above lemma. For instance, the space $c_{00}$ is a dense subspace of the space $c_0$, as was shown in Section 4.1.

Let $X$ be an infinite-dimensional normed space with a countable Hamel basis $e_1, e_2, \ldots$. Each of the spaces

$$X_n = \mathrm{span}\{e_1, \ldots, e_n\}, \qquad n = 1, 2, \ldots,$$

is finite-dimensional and therefore is nowhere dense in $X$ (cf. Lemma 4.3). It is clear that

$$X = \bigcup_{k=1}^{\infty} X_k.$$

By Baire's theorem (cf. Theorem 2.17), $X$ is not complete. Hence we have the following result.

**Theorem 4.15.** *A Hamel basis of a Banach space is either finite or uncountable.*

## 4.4 Compactness in Normed Spaces

By the Bolzano–Weierstrass theorem, the closed unit ball

$$\overline{B}(0,1) = \{x \in \ell_2^n : \|x\| \leq 1\}$$

in the Euclidean space $\ell_2^n$ is a compact set (cf. Section 2.7). This is no longer true for the space $\ell_2$, which is the closest infinite-dimensional analogue of $\ell_2^n$. Indeed, consider the Schauder basis $(e_n)$ in $\ell_2$ defined by $e_n = (\delta_{nk})$, where

$$\delta_{nk} = \begin{cases} 0, & \text{if } k \neq n, \\ 1, & \text{if } k = n, \end{cases}$$

is the *Kronecker delta* (cf. Exercise 4.3). It is clear that for $k \neq n$, we have $d(e_k, e_n) = \sqrt{2}$. It follows that $(e_n)$ has no Cauchy subsequences and therefore no convergent subsequences. Hence the closed unit ball $\overline{B}(0,1)$ in $\ell_2$ is not compact.

It turns out that the closed unit ball $\overline{B}(0,1)$ of every infinite-dimensional normed space $X$ is a noncompact set. Following the idea used in the case of $\ell_2$, we want to construct a sequence $(e_n)$ of unit vectors in $X$ such that $d(e_k, e_n) > a$ for all $k \neq n$, where $a$ is a positive number.

For this we need the result known as *Riesz's lemma*.

**Theorem 4.16.** *Let $X_0$ be a proper closed subspace of a normed space $X$. Then for each $\delta \in (0,1)$, there existts $x_\delta \in X$ such that $\|x_\delta\| = 1$ and*

$$\|x_\delta - y\| \geq \delta, \qquad \text{for all } y \in X_0.$$

*Proof.* Select $x_1 \in X \setminus X_0$ with $\|x_1\| = 1$, and let

$$d = \inf_{y \in X_0} \|x_1 - y\|$$

be the distance from $x_1$ to the subspace $X_0$. Note that $d > 0$, because $X_0$ is a closed subspace of $X$. Because $d/\delta > d$, we can choose $x_0 \in X_0$ such that $\|x_1 - x_0\| < d/\delta$. We define

$$x_\delta = \frac{x_1 - x_0}{\|x_1 - x_0\|},$$

so $\|x_\delta\| = 1$ (cf. Figure 4.2). Then for every $y \in X_0$, we have

$$\|x_\delta - y\| = \left\| \frac{x_1 - x_0}{\|x_1 - x_0\|} - y \right\| = \frac{1}{\|x_1 - x_0\|} \left\| x_1 - (x_0 + \|x_1 - x_0\| y) \right\| \geq \frac{\delta}{d} d = \delta,$$

because $x_0 + \|x_1 - x_0\| y \in X_0$. $\qquad\square$

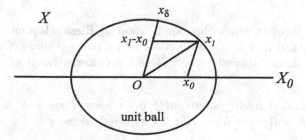

**Fig. 4.2** A hint for the proof of Riesz's lemma.

We apply Riesz's lemma to establish the desired result.

**Theorem 4.17.** *The closed unit ball in a normed space $X$ is compact if and only if $X$ is finite-dimensional.*

*Proof.* (Necessity.) Suppose that the closed unit ball $\overline{B}(0,1)$ is compact but the space is not finite-dimensional. Let $x_1$ be a unit vector in $X$. The 1-dimensional subspace of $X$ generated by $x_1$ is closed (cf. Theorem 4.13). By Riesz's lemma, there is a unit vector $x_2$ such that

$$\|x_2 - x_1\| \geq \delta = \frac{1}{2}.$$

Suppose we constructed a set $\{x_1, \ldots, x_n\}$ of unit vectors in $X$ such that

$$\|x_p - x_q\| \geq \frac{1}{2}, \qquad \text{for } p, q \leq n, \; p \neq q.$$

The subspace of $X$ generated by these vectors is finite-dimensional and therefore closed (cf. Theorem 4.13). Hence by Riesz's lemma, there exists a unit vector $x_{n+1}$ such that

$$\|x_p - x_{n+1}\| \geq \frac{1}{2}, \qquad \text{for all } p \leq n.$$

By induction, we obtain an infinite sequence $(x_n)$ of unit vectors in $X$ with the property

$$\|x_p - x_q\| \geq \frac{1}{2}, \qquad \text{for all } p \neq q.$$

Clearly, this sequence cannot have a convergent subsequence, which contradicts compactness of the unit ball. Therefore, $X$ is finite-dimensional.

(Sufficiency.) Let $B = \overline{B}(0,1)$ be the closed unit ball in an $n$-dimensional normed space $X$, and $T : X \to \ell_2^n$ an isomorphism (cf. Theorem 4.12). The set $T(B)$ is bounded because $T$ is bounded, and closed because $T^{-1}$ is continuous $(T(B) = (T^{-1})^{-1}(B))$. Hence by the Bolzano–Weierstrass theorem, $T(B)$ is compact. By Theorem 2.20, the image of a compact set under a continuous mapping is compact. Because $B = T^{-1}(T(B))$ and $T^{-1}$ is continuous, $B$ is compact. □

The next theorem is another application of Riesz's lemma. We proved (cf. Theorem 4.6) that a linear functional on a normed space $X$ is bounded if and only if its null space is closed. Here is an alternative proof of the same theorem.

**Theorem 4.18.** *A linear functional $f$ on a normed space $X$ is bounded if and only if its null space $X_0 = \mathcal{N}(f)$ is a closed subspace of $X$.*

*Proof.* (Necessity.) The theorem is a special case of Theorem 4.1 (for bounded operators).

(Sufficiency.) We may assume that $f$ is a nonzero functional on $X$. Suppose that $X_0 = \mathcal{N}(f)$ is a proper closed subspace of $X$. By Riesz's lemma, there exists $x \in X \setminus X_0$ such that $\|x - y\| \geq 1/2$ for all $y \in X_0$. Every vector $z \in X \setminus X_0$ can be written in the form

$$z = \alpha(x - y), \qquad \text{where } \alpha \neq 0 \text{ and } y \in X_0$$

(cf. Exercise 4.19). We have

$$\frac{|f(z)|}{\|z\|} = \frac{|\alpha||f(x)|}{|\alpha|\|x - y\|} \leq 2|f(x)|.$$

This inequality also holds for every nonzero vector $z \in X_0$. Hence $f$ is a bounded functional on $X$. $\qquad\square$

Riesz's lemma states that if $X_0$ is a proper closed subspace of $X$, then there exist points on the surface of the unit ball whose distance $d$ from $X_0$ is as close to 1 as desired. Note that the distance $d$ cannot be greater than 1, because for a unit vector $x$, one has

$$d = \inf_{y \in X_0} \|x - y\| \leq \|x - 0\| = 1.$$

However, it need not be true that there are points on the unit sphere that are at distance 1 from $X_0$. To support this statement, we introduce geometric concepts motivated by properties of perpendicular lines in Euclidean geometry.

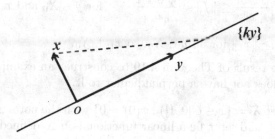

**Fig. 4.3** $x$ is orthogonal to $y$.

Let $x$ and $y$ be nonzero vectors in a normed space $X$. We say that $x$ is *orthogonal* to $y$, and write $x \perp y$, if $\|x - ky\| \geq \|x\|$ for all scalars $k$. In geometry, $x$ is orthogonal to $y$ if $\|x\|$ is the shortest distance from $x$ to the 1-dimensional subspace $\{ky : k \in \mathbf{F}\}$ of the space $X$ (cf. Fig. 4.3).

Let $M$ be a subspace of $X$. We say that a vector $x \in X \setminus M$ is *orthogonal* to $M$, and write $x \perp M$, if $x \perp y$ for all $y \in M$. It is clear that $x \perp M$ if and only if $\|x - y\| \geq \|x\|$ for all $y \in M$. If such a vector $x \in X \setminus M$ exists and $\|x\| = 1$, then we say that $x$ is a *perpendicular* to $M$.

The next theorem characterizes perpendiculars to null spaces of bounded linear functionals.

**Theorem 4.19.** *Let $X$ be a normed space, $f$ a bounded nonzero linear functional on $X$, and $X_0$ the "hyperplane"*

$$X_0 = \mathcal{N}(f) = \{y \in X : f(y) = 0\}.$$

*A vector $x_1 \in X \setminus X_0$ is orthogonal to $X_0$ if and only if $|f(x_1)|/\|x_1\| = \|f\|$.*

*Proof.* (Necessity.) Let $x_1$ be a vector in $X \setminus X_0$. Every vector $x \in X \setminus X_0$ can be written in the form

$$x = \alpha(x_1 - y), \qquad \text{where } \alpha \neq 0 \text{ and } y \in X_0$$

(cf. Exercise 4.19). We have

$$\frac{|f(x)|}{\|x\|} = \frac{|\alpha||f(x_1)|}{|\alpha|\|x_1 - y\|} = \frac{|f(x_1)|}{\|x_1 - y\|}. \tag{4.2}$$

If $x_1 \perp X_0$, that is, $\|x_1 - y\| \geq \|x_1\|$ for all $y \in X_0$, then by (4.2), we have

$$\frac{|f(x)|}{\|x\|} = \frac{|f(x_1)|}{\|x_1 - y\|} \leq \frac{|f(x_1)|}{\|x_1\|},$$

so $\|f\| = \sup_{x \in X \setminus X_0}\{|f(x)|/\|x\|\} = |f(x_1)|/\|x_1\|$.
(Sufficiency.) If $\|f\| = |f(x_1)|/\|x_1\|$, then again by (4.2), we have

$$\frac{|f(x_1)|}{\|x_1 - y\|} = \frac{|f(x)|}{\|x\|} \leq \|f\| = \frac{|f(x_1)|}{\|x_1\|}, \qquad \text{for } y \in X_0 \text{ and } x = x_1 - y.$$

Hence $\|x_1 - y\| \geq \|x_1\|$ for all $y \in X_0$, so $x_1 \perp X_0$.  $\square$

We apply the result of Theorem 4.19 to construct an example of a closed subspace that does not have a perpendicular to it.

*Example 4.9.* Let $X = \{x \in C[0,1]) : x(0) = 0\}$ with the norm inherited from the space $C[0,1]$, and let $f$ be a linear functional on $X$ defined by

$$f(x) = \int_0^1 x(t)\,dt.$$

The norm of this functional is 1 (cf. Exercise 4.20). Let $x_1$ be a unit vector in $X \setminus X_0$, where $X_0$ is defined as in Theorem 4.19. We have

$$|f(x_1)| = \left|\int_0^1 x_1(t)\,dt\right| \leq \int_0^1 |x_1(t)|\,dt < 1,$$

because $\|x_1\| = \max_{t \in [0,1]} |x_1(t)| = 1$ and $x_1(0) = 0$ (cf. Exercise 4.21). By Theorem 4.19, the properties $\|f\| = 1$ and $|f(x_1)| < 1$ imply that $x_1$ is not a perpendicular to $X_0$.

# Notes

As they are in real and complex analysis, convergent and absolutely convergent series are important tools in functional analysis. For instance, in this book, we use series to prove the famous open mapping theorem in Chapter 6.

There are critical differences between the concepts of Hamel and Schauder basis. An obvious one is clearly that a Schauder basis is a sequence of vectors, whereas a Hamel basis is just a set of vectors in the space.

According to Theorem 4.1, a normed space with a Schauder basis is separable. However, as was shown in 1973 by Per Enflo (*Acta Mathematica* **130**, pp. 309–317), there are separable spaces without Schauder basis. Of course, these spaces have Hamel bases.

The operator of indefinite integration introduced in Example 4.6 on the space $C[0,1]$ can be defined by the same formula on the normed space $L_2[0,1]$ (the space of Lebesgue measurable functions $x$ on $[0,1]$ such that the Lebesgue integral $\int_0^1 |x(t)|^2 \, dt$ is bounded and with the norm $\|x\| = \sqrt{\int_0^1 |x(t)|^2 \, dt}$). In this case, this operator is known as the *Volterra operator*. The norm of the Volterra operator is $2/\pi$ (cf. Example 7.3 in Chapter 7 and Problem 148 in Halmos 1967).

By Theorem 4.12, every two $n$-dimensional normed spaces are isomorphic. In fact, every two $n$-dimensional metric vector spaces are isomorphic, that is, they are isomorphic as vector spaces and homeomorphic as metric spaces. Specifically, all spaces $\ell_p^n$, $p \in (0, \infty]$ (cf. Section 3.3), are isomorphic (cf. Exercise 4.17). Moreover, every two Hausdorff topological vector spaces of the same finite dimension are isomorphic (Rudin 1973, Theorem 1.21).

The multiplication operators from Example 4.7 are, in a natural sense, the canonical examples of normal operators. This is established in the spectral theory in Hilbert spaces, which is beyond the scope of this book.

# Exercises

**4.1.** Give an example of an absolutely convergent series in a normed space $X$ that is not convergent. (Hint: Try $X = c_{00}$.)

**4.2.** (a) If in a normed space $X$, the absolute convergence of every series always implies convergence of the series, show that $X$ is a Banach space.

(b) Show that in a Banach space, an absolutely convergent series is convergent.

**4.3.** Show that vectors $(e_n)$, where $e_n$ is the sequence whose $n$th term is 1 and all other terms are zero,

$$e_1 = (1, 0, 0, \ldots),$$
$$e_2 = (0, 1, 0, \ldots),$$
$$\ldots$$

form a Schauder basis in $\ell_p$ for every $p \in [1, \infty)$, and in the spaces $c_0$ and $c_{00}$.

**4.4.** Prove that a mapping from a metric space $X$ to a metric space $Y$ is continuous if and only if the inverse image of every closed subset of $Y$ is closed in $X$ (cf. Theorem 2.11).

**4.5.** Show that a linear operator is bounded if and only if it maps bounded sets in $X$ into bounded subsets of $Y$.

**4.6.** Prove Theorem 4.5.

**4.7.** Show that the integral operator $T$ in Example 4.3 is linear and maps $C[0, 1]$ to $C[0, 1]$.

**4.8.** Let $f$ and $g$ be two bounded real-valued functions on the set $A \subseteq \mathbf{R}$. Show that
$$\sup_{x \in A}(f(x) + g(x)) \leq \sup_{x \in A} f(x) + \sup_{x \in A} g(x).$$

**4.9.** Prove that operators of the left and right shift on $\ell_p$ (cf. Example 4.4) are bounded and $\|T_l\| = \|T_r\| = 1$.

**4.10. (Bounded linear extension theorem)** Let $X$ be a normed space and $\widetilde{X}$ its Banach completion. If $f$ is a bounded linear functional on $X$, then there exists a unique linear functional $\widetilde{f}$ on $\widetilde{X}$ such that $\widetilde{f}|_X = f$ and $\|\widetilde{f}\| = \|f\|$.

**4.11.** Let $X$, $Y$, and $Z$ be normed spaces and $S : X \to Y$ and $T : Y \to Z$ bounded operators. Prove that

$$\|TS\| \leq \|T\|\|S\|.$$

**4.12.** Let $T$ be a bounded operator from a normed space $X$ to a normed space $Y$. Prove that for every $x \in X$ and $r > 0$,

$$\sup_{y \in B(x,r)} \|Ty\| \geq \|T\| r.$$

**4.13.** Show that for nonzero $a$ and $b$, the function $\sqrt{x_1^2/a^2 + x_2^2/b^2}$ defines a norm on the 2-dimensional real vector space.

**4.14.** Show that the norms on $\mathbf{R}^2$ defined by

$$\|x\| = \sqrt{x_1^2 + x_2^2} \qquad \text{and} \qquad \|x\|' = \sqrt{x_1^2/a^2 + x_2^2/b^2},$$

for $a > b > 0$, satisfy the inequalities

$$\frac{1}{a}\|x\| \leq \|x\|' \leq \frac{1}{b}\|x\|, \qquad \text{for all } x \in \mathbf{R}^2.$$

**4.15.** Prove that the set

$$S = \{(c_1, \ldots, c_n) : \sum_k |c_k| = 1\}$$

is closed and bounded in $\ell_2^n$.

**4.16.** Show that if $T$ is an isomorphism of normed spaces $X$ and $Y$, then both $T$ and $T^{-1}$ are bounded operators.

**4.17.** Show that every space $\ell_p^n$ for $0 < p < 1$ is isomorphic to $\ell_\infty^n$. Conclude that all spaces $\ell_p^n$, $p \in (0, \infty]$, are isomorphic.

**4.18.** Show that two equivalent norms (cf. Definition 4.4) on a vector space define equivalent operator norms.

**4.19.** Let $X_0$ be a proper subspace of the vector space $X$ over the field $\mathbf{F}$ and $x \in X \setminus X_0$. Show that every vector $z \in X \setminus X_0$ can be written in the form

$$z = \alpha(x - y), \qquad \text{where } \alpha \neq 0 \text{ and } y \in X_0.$$

**4.20.** Show that $\|f\| = 1$ for the functional $f$ from Example 4.9.

**4.21.** Let $x$ be a continuous nonnegative function on $[0, 1]$ such that $x(0) = 0$ and $x(t) \leq 1$ for all $t \in [0, 1]$. Show that $\int_0^1 x(t)\, dt < 1$.

# Chapter 5
# Linear Functionals

The concept of a linear functional is one of the main concepts in linear algebra and one that plays a significant role in functional analysis. In this chapter, we present basic facts of the theory of linear functionals on topological vector spaces. First, in Section 5.1, we describe bounded the linear functionals on sequence spaces introduced in Chapter 3. These results show that the dual spaces of these normed spaces are very rich. The natural question about the mere existence of nontrivial linear functionals is addressed by various Hahn–Banach theorems, which are found in Sections 5.2 and 5.3. An application of Hahn–Banach theorems is presented in Section 5.4, where all bounded linear functionals on the space $C[a, b]$ are described. Finally, reflexive spaces and their elementary properties are introduced in Section 5.5.

## 5.1 Dual Spaces of Sequence Spaces

In this section, we determine the dual spaces of the sequence spaces introduced in Chapter 3.

Let $(e_n)$ be a Schauder basis in a normed space $(X, \| \cdot \|)$. Every vector $x \in X$ admits a unique expansion

$$x = \sum_{k=1}^{\infty} x_k e_k,$$

where each $x_k$ is a (linear) function of $x$, that is, $x_k = x_k(x)$ (cf. Exercise 5.2). The correspondence

$$x \mapsto (x_1(x), \ldots, x_n(x), \ldots)$$

defines an isomorphism from the vector space $X$ onto a sequence vector space consisting of all sequences $(x_1(x), \ldots, x_n(x), \ldots)$ for $x \in X$. If we define

© Springer International Publishing AG, part of Springer Nature 2018
S. Ovchinnikov, *Functional Analysis*, Universitext,
https://doi.org/10.1007/978-3-319-91512-8_5

$$\|(x_1(x), \dots, x_n(x), \dots)\| = \|x\|, \qquad \text{for } x \in X,$$

then the normed space $X$ is isomorphic to a normed sequence space.

Let $f$ be a bounded linear functional on $X$, so $f \in X^*$, and $b_k = f(e_k)$ for $k \in \mathbf{N}$. Because $f$ is a continuous function, we have, for $x = \sum_{k=1}^{\infty} x_k e_k$,

$$f(x) = \sum_{k=1}^{\infty} x_k f(e_k) = \sum_{k=1}^{\infty} x_k b_k, \qquad \text{for } x \in X. \tag{5.1}$$

Hence $f$ is completely determined by the sequence $(b_1, \dots, b_n, \dots)$. In other words, the dual vector space $X^*$ can be identified with a vector sequence space

$$\{(b_1, \dots, b_n, \dots)\}, \qquad \text{for } f \in X^*.$$

These observations justify our interest in normed sequence spaces in the context of the duality problem. For this, we use formula (5.1) frequently.

We begin with the relatively simple case of the space $\ell_1$. Let $(e_n)$ be the standard Schauder basis in $\ell_1$ (cf. Exercise 4.3) and $x = \sum_{k=1}^{\infty} x_k e_k$ the expansion of $x \in \ell_1$, so $x = (x_1, \dots, x_n, \dots)$ with $\|x\|_1 = \sum_{k=1}^{\infty} |x_k|$.

For a bounded linear functional $f$ on $\ell_1$, let $b_n = f(e_n)$ for $n \in \mathbf{N}$. Then

$$|b_n| = |f(e_n)| \leq \|f\| \|e_n\| = \|f\|,$$

so $(b_n)$ is a bounded sequence. Hence $(b_n) \in \ell_\infty$ with

$$\|(b_n)\|_\infty = \sup_{n \in \mathbf{N}} |b_n| \leq \|f\|. \tag{5.2}$$

On the other hand, for every $(b_n) \in \ell_\infty$, we may define a bounded linear functional $f$ on $\ell_1$ by (5.1) as follows:

$$f(x) = \sum_{k=1}^{\infty} x_k b_k, \qquad \text{for } x = (x_1, \dots, x_n, \dots) \in \ell_1.$$

Clearly, $f$ is linear and

$$|f(x)| = \left| \sum_{k=1}^{\infty} x_k b_k \right| \leq \sup_{k \in \mathbf{N}} |b_k| \|x\|_1 = \|(b_n)\|_\infty \|x\|_1, \tag{5.3}$$

so $f$ is indeed a bounded linear functional on $\ell_1$, that is, $f \in \ell_1^*$.

The correspondence $T : f \mapsto (b_n)$ is an isomorphism of vector spaces $\ell_1^*$ and $\ell_\infty$. By (5.3), $\|f\| \leq \|(b_n)\|_\infty$, which implies together with (5.2) that $\|f\| = \|(b_n)\|_\infty$. Hence $T$ is an isomorphism of $\ell_1^*$ onto $\ell_\infty$, so we can write $\ell_1^* = \ell_\infty$. We have proved the following theorem.

**Theorem 5.1.** *The dual of the space $\ell_1$ is the space $\ell_\infty$.*

Consider now the space $c_0$ of all sequences converging to zero. Let $(e_n)$ be the standard Schauder basis in $c_0$ (cf. Exercise 4.3) and $f$ a bounded linear functional on $c_0$. Let $x = \sum_{k=1}^{\infty} x_k e_k$ be the expansion of $x \in c_0$, so again by (5.1),

$$f(x) = \sum_{k=1}^{\infty} x_k b_k.$$

For $n \in \mathbf{N}$, we define a sequence $x^{(n)} \in c_0$ by

$$x^{(n)} = (x_1, \ldots, x_k, \ldots, x_n, 0, 0, \ldots),$$

where for $1 \le k \le n$,

$$x_k = \begin{cases} 1, & \text{if } b_k = 0, \\ |b_k|/b_k, & \text{otherwise.} \end{cases}$$

Note that $x_k b_k = |b_k|$ for $1 \le k \le n$. We have

$$\|f\| \|x^{(n)}\|_{\infty} \ge |f(x^{(n)})| = \left| \sum_{k=1}^{n} x_k b_k \right| = \sum_{k=1}^{n} |b_k|,$$

for every $n \in \mathbf{N}$. Hence $\sum_{k=1}^{\infty} |b_k| < \infty$, that is, $(b_n) \in \ell_1$. Moreover, it follows that

$$\|(b_n)\|_1 \le \|f\|, \tag{5.4}$$

because $\|x^{(n)}\|_{\infty} = 1$.

For $(b_n) \in \ell_1$ and $x = \sum_{k=1}^{\infty} x_k e_k \in c_0$, we have $|x_k b_k| \le \|x\|_{\infty} |b_k|$ for all $k \in \mathbf{N}$. Then

$$\sum_{k=1}^{\infty} |x_k b_k| \le \|x\|_{\infty} \sum_{k=1}^{\infty} |b_k| = \|x\|_{\infty} \|(b_n)\|_1.$$

Hence the series $\sum_{k=1}^{\infty} x_k b_k$ converges absolutely, and therefore the function

$$f(x) = \sum_{k=1}^{\infty} x_k b_k$$

is well defined. Clearly, $f$ is a linear functional on $(c_0)$. Furthermore,

$$|f(x)| \le \sum_{k=1}^{\infty} |x_k| |b_k| \le \|x\|_{\infty} \|(b_n)\|_1,$$

so $\|f\| \le \|(b_n)\|_1$. By (5.4), $\|f\| = \|(b_n)\|_1$. Therefore, the correspondence $T : f \mapsto (b_n)$ is an isomorphism of the normed spaces $c_0^*$ and $\ell_1$.

**Theorem 5.2.** *The dual of the space $c_0$ is the space $\ell_1$.*

We leave the proof of the next theorem as an exercise (cf. Exercise 5.3).

**Theorem 5.3.** *The dual of the space $c$ is the space $\ell_1$.*

Note that the two different spaces $c_0$ and $c$ have the same duals. Incidentally, the dual space of $c_{00}$ is also $\ell_1$. Indeed, by Exercise 3.21(a), $c_{00}$ is dense in $c_0$. Hence by the "bounded linear extension theorem" (cf. Exercise 4.10), a bounded linear functional on $c_{00}$ has a unique extension to a bounded linear functional on $c_0$ with the same norm.

Now we determine the dual space of the space $\ell_p$ for $p \in (0, \infty)$. In what follows, $q$ is the conjugate exponent of $p$, that is,

$$\frac{1}{p} + \frac{1}{q} = 1.$$

We will need the obvious identity $p(q-1) = q$.

As before, let $f$ be a bounded linear functional on $\ell_p$. For $x = \sum_{k=1}^{\infty} x_k e_k$ in $\ell_p$ (here $(e_n)$ is the Schauder basis in $\ell_p$ as in Exercise 4.3), we have, by (5.1),

$$f(x) = \sum_{k=1}^{\infty} x_k b_k, \tag{5.5}$$

where $b_k = f(e_k)$ for $k \in \mathbf{N}$. We need to describe the sequence $(b_k)$ that completely determines $f$. For $n \in \mathbf{N}$, define a sequence $x^{(n)} \in \ell_p$ by

$$x^{(n)} = (x_1, \ldots, x_k, \ldots, x_n, 0, 0, \ldots),$$

where for $1 \leq k \leq n$, $x_k$ is chosen to satisfy $|x_k b_k| = |b_k|^q$. For this, we define

$$x_k = \begin{cases} 0, & \text{if } b_k = 0, \\ |b_k|^q / b_k, & \text{otherwise.} \end{cases}$$

Then

$$f(x^{(n)}) = \sum_{k=1}^{n} |b_k|^q.$$

On the other hand, because $|x_k| = |b_k|^{q-1}$, we have

$$\sum_{k=1}^{n} |b_k|^q = f(x^{(n)}) = |f(x^{(n)})| \leq \|f\| \|x^{(n)}\|_p = \|f\| \left( \sum_{k=1}^{n} |x_k|^p \right)^{1/p}$$

$$= \|f\| \left( \sum_{k=1}^{n} |b_k|^{p(q-1)} \right)^{1/p} = \|f\| \left( \sum_{k=1}^{n} |b_k|^q \right)^{1/p}.$$

From this we immediately obtain (assuming that not all the $b_k$ are zero)

$$\left(\sum_{k=1}^{n}|b_k|^q\right)^{1-1/p} = \left(\sum_{k=1}^{n}|b_k|^q\right)^{1/q} \leq \|f\|.$$

This inequality holds trivially if $b_k = 0$ for $1 \leq k \leq n$. Hence

$$\|(b_n)\|_q = \left(\sum_{k=1}^{\infty}|b_k|^q\right)^{1/q} \leq \|f\|, \qquad (5.6)$$

so $(b_n) \in \ell_q$.

On the other hand, for an arbitrary sequence $(b_n) \in \ell_q$, formula (5.5) defines a bounded linear functional on the space $\ell_p$. Indeed, the linearity of $f$ is obvious, and by Hölder's inequality,

$$|f(x)| = \left|\sum_{k=1}^{\infty} x_k b_k\right| \leq \sum_{k=1}^{\infty} |x_k b_k| \leq \|x\|_p \|(b_k)\|_q, \qquad (5.7)$$

so $f$ is bounded.

We have shown that formula (5.5) is the general form of a bounded linear functional on $\ell_p$. It is clear that $T : f \mapsto (b_n)$ is an isomorphism of the vector space $\ell_p^*$ onto the vector space $\ell_q$. It follows from (5.6) and (5.7) that $\|f\| = \|(b_n)\|_q$, so in fact, $T$ is an isomorphism of the normed spaces $\ell_p^*$ and $\ell_q$.

**Theorem 5.4.** *The dual of the space $\ell_p$ is the space $\ell_q$, where $p$ and $q$ are conjugate exponents.*

**Corollary 5.1.** *The dual space of $\ell_2$ is the space $\ell_2$ itself.*

We extend the concept of a dual space to arbitrary topological vector spaces by defining the dual space $X^*$ of a topological vector space $X$ as the set of all continuous linear functionals on $X$. It is clear that $X^*$ is a vector space over the field $\mathbf{F}$.

In the rest of this section, we describe the dual of the metric vector space $s$ of all sequences with the distance function defined by (cf. Section 3.5)

$$d(x,y) = \sum_{k=1}^{\infty} \frac{1}{2^k}\frac{|x_k - y_k|}{1 + |x_k - y_k|}, \qquad \text{for } (x_n), (y_n) \in s.$$

For $x = (x_1, x_2, \ldots)$ we define $x^{(n)} = (x_1, \ldots, x_n, 0, 0, \ldots)$. Then

$$d(x^{(n)}, x) = \sum_{k=n+1}^{\infty} \frac{1}{2^k}\frac{|x_k|}{1+|x_k|} \leq \sum_{k=n+1}^{\infty}\frac{1}{2^k} = \frac{1}{2^n} \to 0, \qquad \text{as } n \to \infty.$$

Hence $x^{(n)} \to x$ in $s$.

Let $f$ be a continuous linear functional on $s$. As before, for $n \in \mathbf{N}$ we denote by $e_n$ the vector $(x_1, \ldots, x_n, \ldots) \in s$ such that $x_n = 1$ and $x_m = 0$

for $m \neq n$, and set $b_n = f(e_n)$. It is clear that $x^{(n)} = \sum_{k=1}^{n} x_k e_k$. Because $f$ is continuous, we have

$$f(x) = \lim_{n \to \infty} f(x^{(n)}) = \lim_{n \to \infty} \sum_{k=1}^{n} x_k f(e_k) = \sum_{k=1}^{\infty} x_k b_k, \qquad (5.8)$$

so the series on the right-hand side converges for every sequence $(x_n)$, and the functional $f$ is completely determined by the sequence $(b_n)$. For every $b_n \neq 0$, we choose $x_n = 1/b_n$. Because $\sum_{k=1}^{\infty} x_k b_k$ converges, it follows that $(b_n)$ contains only a finite number of nonzero terms. Hence $(b_n) \in c_{00}$. On the other hand, every sequence $(b_n) \in c_{00}$ defines, by (5.8), a continuous linear functional on $s$ (cf. Exercise 5.4).

**Theorem 5.5.** *The dual of the metric vector space $s$ is the vector space $c_{00}$.*

Note that the space $c_{00}$ endowed with the sup-norm is not a Banach space, unlike the duals of normed spaces. Also note that the sequence $(e_n)$ may be considered an analogue of the Schauder basis for the (unnormable) sequence space $s$.

## 5.2 Hahn–Banach Theorems

As we already know (cf. Section 5.1), there are plenty of continuous linear functionals on some normed spaces. One can wonder whether that is true for an arbitrary topological vector space.

First we address this issue from a purely algebraic standpoint. On a finite-dimensional vector space over the field $\mathbf{F}$, all linear functionals are effectively described. For an infinite-dimensional vector space, it is natural to consider an extension problem. Namely, if $f$ is a nontrivial linear functional on a proper subspace $V_0$ of a vector space $V$, determine whether there is a linear functional $\tilde{f}$ on $V$ such that its restriction to $V_0$ is $f$. If such a functional $\tilde{f}$ exists, it is called an *extension* of $f$ from $V_0$ onto $V$. It is not difficult to see that $\tilde{f}$ always exists. For this, note that every $z \in V$ can be uniquely written as $z = x + y$, where $x \in V_0$ and $y \in V_1$, the algebraic complement of $V_0$ in $V$. Then one defines $\tilde{f}(z) = f(x)$.

Suppose that $X$ is a topological vector space over $\mathbf{F}$. Is there a nontrivial continuous linear functional on $X$? In general, the answer is negative, as we will see later (cf. Example 5.4). However, the answer is affirmative in the case of seminormed (and therefore normed) spaces, as the Hahn–Banach Theorems 5.6 and 5.7 assert.

Let $X$ be a vector space endowed with a seminorm $p$. A linear functional $f$ on $X$ is said to be *dominated* by $p$ if

$$|f(x)| \leq p(x), \qquad \text{for all } x \in X.$$

If $f$ is dominated by a seminorm $p$, then it is necessarily continuous in the topology defined by $p$. Indeed, for $x_0 \in X$ and $\varepsilon > 0$, if

$$x \in B(x_0, \varepsilon) = \{x \in X : p(x - x_0) < \varepsilon\},$$

then

$$|f(x) - f(x_0)| = |f(x - x_0)| \leq p(x - x_0) < \varepsilon.$$

The following is a version of the Hahn–Banach theorem in the case of the field of real numbers, $\mathbf{F} = \mathbf{R}$. A more general version will be established later (Theorem 5.8).

**Theorem 5.6.** *Suppose $(X, p)$ is a real seminormed space and $Y$ a subspace of $X$. If $f$ is a real linear functional on $Y$ dominated by $p$, so that*

$$|f(x)| \leq p(x), \qquad \text{for all } x \in Y,$$

*then there is a linear functional $\widetilde{f}$ on $X$ that is an extension of $f$ and is dominated by $p$, that is,*

$$\widetilde{f}(x) = f(x), \qquad \text{for all } x \in Y,$$

*and*

$$|\widetilde{f}(x)| \leq p(x), \qquad \text{for all } x \in X.$$

Before proving Theorem 5.6, we prove a special case of it.

**Lemma 5.1.** *Suppose $X$ is a seminormed space with the seminorm $p$, $Y$ a proper subspace of $X$, $x_0$ a vector in $X \setminus Y$, and $Z = \mathrm{span}(x_0, Y)$.*
*If $f$ is a linear functional on $Y$ dominated by $p$, so that*

$$|f(x)| \leq p(x), \qquad \text{for all } x \in Y,$$

*then there is a linear functional $\widetilde{f}$ on $Z$ that is an extension of $f$ and dominated by $p$, that is,*

$$\widetilde{f}(x) = f(x), \qquad \text{for all } x \in Y,$$

*and*

$$|\widetilde{f}(x)| \leq p(x), \qquad \text{for all } x \in Z.$$

*Proof.* Inasmuch as $Z = \mathrm{span}(x_0, Y)$, for every vector $x \in Z$ there exist $\lambda \in \mathbf{R}$ and $y \in Y$ such that

$$x = \lambda x_0 + y.$$

Let $\widetilde{f}$ be a linear functional on $Z$. Then

$$\widetilde{f}(x) = \lambda \widetilde{f}(x_0) + \widetilde{f}(y).$$

We want $\widetilde{f}$ to be an extension of $f$, so $\widetilde{f}(y) = f(y)$ for all $y \in Y$, and $\widetilde{f}$ should be dominated by $p$ on $Z$, so

$$|\widetilde{f}(x)| = |\lambda \widetilde{f}(x_0) + f(y)| \le p(x) = p(\lambda x_0 + y), \qquad \text{for all } \lambda \in \mathbf{R} \text{ and } y \in Y.$$

Because $f$ is dominated by $p$ on $Y$, the above inequality holds for $\lambda = 0$; hence we assume that $\lambda \ne 0$. By setting $y = -\lambda y_1$ and $c = \widetilde{f}(x_0)$, we may write the desired inequality as

$$|\lambda c - \lambda f(y_1)| \le p(\lambda x_0 - \lambda y_1),$$

or dividing by $|\lambda|$ and expanding, as

$$-p(x_0 - y_1) + f(y_1) \le c \le p(x_0 - y_1) + f(y_1),$$

which we want to hold for all $y_1 \in Y$. To show that there exists $c$ satisfying all these inequalities, it suffices to show that

$$-p(x_0 - y') + f(y') \le p(x_0 - y'') + f(y''), \qquad \text{for all } y', y'' \in Y.$$

This follows immediately from

$$\begin{aligned} f(y') - f(y'') = f(y' - y'') &\le p(y' - y'') = p((y' - x_0) + (x_0 - y'')) \\ &\le p(y' - x_0) + p(x_0 - y'') = p(x_0 - y') + p(x_0 - y''), \end{aligned}$$

for all $y', y'' \in Y$.                                                                 $\square$

Now we proceed with the proof of Theorem 5.6.

*Proof.* Let $\mathcal{Y}$ be the collection of all pairs $(Y_1, f_1)$, where $Y_1$ is a subspace of $X$ containing $Y$ and $f_1$ is a linear functional on $Y_1$ dominated by $p$ and coinciding with $f$ on $Y$. The set $\mathcal{Y}$ is not empty, because it contains the pair $(Y, f)$. We define a partial order $\preceq$ on $\mathcal{Y}$ by

$$(Y_1, f_1) \preceq (Y_2, f_2) \quad \text{if and only if} \quad Y_1 \subseteq Y_2 \text{ and } f_2 \text{ is an extension of } f_1.$$

Every chain $\{(Y_j, f_j)\}_{j \in J}$ in $\mathcal{Y}$ has an upper bound, for instance

$$\left( \bigcup_{j \in J} Y_j, f' \right),$$

where by definition, $f' = f_j$ on $Y_j$. By Zorn's lemma, $\mathcal{Y}$ has a maximal element $(\widehat{Y}, \widehat{f})$. By Lemma 5.1, $\widehat{Y} = X$. Therefore, $\widetilde{f} = \widehat{f}$ is the desired extension of $f$.                                                                 $\square$

Suppose that $X$ is a seminormed space with a nonzero seminorm $p$. Let $x_0$ be a vector in $X$ such that $p(x_0) \ne 0$, and let $X_0$ be a one-dimensional subspace generated by $x_0$. We define $f(tx_0) = kt$, $t \in \mathbf{R}$, where $|k| \le p(x_0)$. Then

$$|f(tx_0)| = |k||t| \le |t|p(x_0) = p(tx_0), \qquad \text{on } X_0.$$

Clearly, $f$ is a linear functional on $X_0$ dominated by $p$. By Theorem 5.6, $f$ admits an extension $\widetilde{f}$ dominated by $p$ onto the entire space $X$. It follows that there are nontrivial continuous functionals on every seminormed space $X$.

The above argument can be used in the case of seminormed complex spaces (cf. Corollary 5.2). For this we need a complex version of the Hahn–Banach theorem.

**Theorem 5.7.** *Suppose $(X, p)$ is a complex seminormed space and $Y$ a subspace of $X$. If $f$ is a complex linear functional on $Y$ dominated by $p$, so that*

$$|f(x)| \le p(x), \qquad \text{for all } x \in Y,$$

*then there is a linear functional $\widetilde{f}$ on $X$ that is an extension of $f$ and dominated by $p$, that is,*

$$\widetilde{f}(x) = f(x), \qquad \text{for all } x \in Y,$$

*and*

$$|\widetilde{f}(x)| \le p(x), \qquad \text{for all } x \in X.$$

*Proof.* We have

$$f(x) = f_1(x) + if_2(x), \qquad x \in Y,$$

where $f_1$ and $f_2$ are the real and imaginary parts of $f$ respectively. Because

$$f_1(ix) + if_2(ix) = f(ix) = if(x) = if_1(x) - f_2(x),$$

we have $f_2(x) = -f_1(ix)$, so

$$f(x) = f_1(x) - if_1(ix), \qquad \text{for } x \in Y.$$

Consider $X$ as a vector space over $\mathbf{R}$. Because

$$|f_1(x)| = |\mathrm{Re}f(x)| \le |f(x)| \le p(x),$$

$f_1$ is dominated on $Y$ by $p$. By Theorem 5.6, $f_1$ has a real extension $\widetilde{f_1}$ on $X$ dominated by $p$. Assuming that $X$ is a complex space, we define

$$\widetilde{f}(x) = \widetilde{f_1}(x) - i\widetilde{f_1}(ix), \qquad \text{for } x \in X.$$

First, we verify that $\widetilde{f}$, so defined, is a complex linear functional on the complex space $X$. Indeed,

$$\widetilde{f}(x+y) = \widetilde{f_1}(x+y) - i\widetilde{f_1}(x+y) = \widetilde{f_1}(x) + \widetilde{f_1}(y) - i[\widetilde{f_1}(ix) + \widetilde{f_1}(iy)]$$
$$= [\widetilde{f_1}(x) - i\widetilde{f_1}(ix)] + [\widetilde{f_1}(y) - i\widetilde{f_1}(iy)] = \widetilde{f}(x) + \widetilde{f}(y),$$

and

$$\widetilde{f}((a+ib)x) = \widetilde{f}_1(ax+ibx) - i\widetilde{f}_1(iax - bx)$$
$$= a\widetilde{f}_1(x) + b\widetilde{f}_1(ix) - i[a\widetilde{f}_1(ix) - b\widetilde{f}_1(x)]$$
$$= (a+ib)[\widetilde{f}_1(x) - i\widetilde{f}_1(ix)] = (a+ib)\widetilde{f}(x).$$

It is clear that $\widetilde{f}$ is an extension of $f$. It remains to show that $|\widetilde{f}(x)| \leq p(x)$ for all $x \in X$. We may assume that $\widetilde{f}(x) \neq 0$. Then for $\mu = |\widetilde{f}(x)|/\widetilde{f}(x)$, we have

$$|\widetilde{f}(x)| = \mu\widetilde{f}(x) = \widetilde{f}(\mu x) = \widetilde{f}_1(\mu x) \leq p(\mu x) = |\mu|p(x) = p(x),$$

because all quantities in the above chain of equalities and inequalities are real, $\widetilde{f}_1$ is dominated by $p$, and $|\mu| = 1$.                                     □

**Corollary 5.2.** *Let $X$ be a seminormed space with a nontrivial seminorm $p$. If $x_0$ is a vector in $X$ such that $p(x_0) > 0$, then there exists a linear functional $f$ on $X$ dominated by $p$ such that $f(x_0) = p(x_0)$.*

*In particular, if $(X, \|\cdot\|)$ is a normed space and $x_0$ a nonzero vector in $X$, then there exists a bounded linear functional $f$ on $X$ such that $\|f\| = 1$ and $f(x_0) = \|x_0\|$.*

*Proof.* Let $X_0 = \{tx_0 : t \in \mathbf{F}\}$ be a 1-dimensional subspace of $X$ generated by $x_0$. Define a functional $g$ on $X_0$ by $g(tx_0) = tp(x_0)$ for $t \in \mathbf{F}$. Then $g$ is a linear functional on $X_0$ dominated by $p$. Theorems 5.6 and 5.7 imply that there is an extension $f$ of $g$ to the space $X$. Hence $f(x_0) = g(x_0) = p(x_0)$.

If $X$ is a normed space, we define $g(tx_0) = t\|x_0\|$ on $X_0 = \{tx_0 : t \in \mathbf{F}\}$. Clearly,

$$|g(tx_0)| = |t|\|x_0\| = \|tx_0\|, \qquad t \in \mathbf{F},$$

so $g$ is dominated by the norm $\|\cdot\|$ and $\|g\| = 1$ on $X_0$. By the Hahn–Banach theorems, there exists an extension $f$ of $g$ dominated by $\|\cdot\|$, that is,

$$f(tx_0) = t\|x_0\| \qquad \text{and} \qquad |f(x)| \leq \|x\|,$$

for all $t \in \mathbf{F}$ and $x \in X$. Then $f(x_0) = \|x_0\|$, and the above inequality implies $\|f\| = 1$. The desired result follows.                                     □

The proof of the following corollary is left as an exercise.

**Corollary 5.3.** *Let $X$ be a normed space, and let $x$ and $y$ be two different vectors in $X$. Then there exists a bounded linear functional $f$ on $X$ such that $f(x) \neq f(y)$.*

**Corollary 5.4.** *Let $Y$ be a closed proper subspace of a seminormed space $X$ and $x_0 \in X \setminus Y$. Then there exists a bounded linear functional $f$ on $X$ such that $f|_Y = 0$ and $f(x_0) \neq 0$.*

*Proof.* Let $X_0 = \{\lambda x_0 + y : \lambda \in \mathbf{F}, \, y \in Y\}$ be the span of the subspace $Y$ and vector $x_0$. Because $x_0 \notin Y$ and $Y$ is closed, there exists $\delta > 0$ such that $\|x_0 + y\| > \delta$ for all $y \in Y$. We define $g(\lambda x_0 + y) = \lambda$. Clearly, $g$ is a linear functional on $X_0$ such that $g|_Y = 0$. For $\lambda \neq 0$, we have

$$\|\lambda x_0 + y\| = |\lambda| \|x_0 + (1/\lambda)y\| \geq |\lambda|\delta.$$

The above inequality also holds for $\lambda = 0$. It follows that

$$|g(\lambda x_0 + y)| = |\lambda| \leq (1/\delta)\|\lambda x_0 + y\|.$$

Thus, $g$ is a bounded linear functional on $Y$. By the Hahn–Banach theorems, it can be extended to a bounded linear functional $f$ on $X$. Furthermore, $f(x_0) = g(x_0) = 1$. $\qquad\square$

The Hahn–Banach theorems claim that a bounded linear functional on a subspace $Y$ of a seminormed space $X$ can be extended to the entire space $X$. Here we address the following question: Is this extension unique? The following three examples illustrate different possibilities.

*Example 5.1.* Let $X = \ell_2^2$, so $p(x, y) = \|(x, y)\| = \sqrt{|x|^2 + |y|^2}$ (the Euclidean plane), $Y = \{(x, 0) : x \in \mathbf{R}\}$, and $f(x) = x$ on $Y$. Clearly, $|f| \leq p$ on $Y$ (in fact, $|f| = p$ on $Y$). Every linear extension $\widetilde{f}$ of $f$ to $X$ is of the form $\widetilde{f}(x, y) = x + \beta y$. The function $\widetilde{f}$ is dominated by $p$ if

$$|x + \beta y| \leq \sqrt{|x|^2 + |y|^2}, \qquad \text{for all } (x, y) \in \mathbf{R}^2.$$

It is not difficult to show (cf. Exercise 5.8) that the above inequality holds for all $x, y \in \mathbf{R}$ only if $\beta = 0$. Hence $\widetilde{f}(x, y) = x$ is the unique extension of the functional $f$ onto the space $X$.

The linear functional $f$ on a proper subspace in the next example has infinitely many bounded extensions to the entire space.

*Example 5.2.* Let $X = \ell_1^2$, so $p(x, y) = \|(x, y)\| = |x| + |y|$, and let $Y$, $f$, and $\widetilde{f}$ be the same as in Example 5.1. For $|\beta| \leq 1$ we have

$$|\widetilde{f}(x, y)| = |x + \beta y| \leq |x| + |\beta||y| < |x| + |y| = p(x, y).$$

Hence all functions $\widetilde{f}(x, y) = |x + \beta y|$ with $|\beta| \leq 1$ are extensions of $f$ that are dominated by $p$ on $X$.

For an infinite-dimensional space we give the following example.

*Example 5.3.* Let $X$ be the space $\ell_\infty$ endowed with the sup-norm, $Y = c$ the subspace of $X$ consisting of all convergent sequences, and $f$ a linear functional on $Y$ defined by $f(x) = \lim x_n$. Clearly, $|f(x)| \leq \|x\| = \sup_{n \in \mathbf{N}} |x_n|$ for all

$x \in Y$. Define subspaces $Y_1$ and $Y_2$ of $X$ as sets of sequences $(x_n)$ such that $(x_{2n})$ and $(x_{2n-1})$ are convergent, respectively. It is clear that $Y \subseteq Y_1 \cap Y_2$. We define linear functionals $f_1$ and $f_2$ on the spaces $Y_1$ and $Y_2$ by

$$f_1(x) = \lim x_{2n} \quad \text{and} \quad f_2(x) = \lim x_{2n-1},$$

respectively. Both functions are dominated by the norm. By the Hahn–Banach theorem, these functions can be extended to linear functionals $\widetilde{f_1}$ and $\widetilde{f_2}$ on $X$. These functionals are also extensions of $f$. They are different extensions because for the sequence $x = ((-1)^n)$ we have

$$\widetilde{f_1}(x) = f_1(x) = 1 \neq -1 = f_2(x) = \widetilde{f_2}(x).$$

We now prove a more general version of Theorem 5.6. A real-valued function $p$ on a real vector space $X$ is said to be a *sublinear functional* if it is *subadditive*, that is,

$$p(x+y) \leq p(x) + p(y), \qquad \text{for all } x,y \in X,$$

and *positive homogeneous*, that is,

$$p(\alpha x) = \alpha p(x), \qquad \text{for all } \alpha \geq 0 \text{ in } \mathbf{R} \text{ and } x \in X.$$

Note that every seminorm is a sublinear functional.

**Theorem 5.8.** *Suppose $X$ is a real vector space and $p$ a sublinear functional on $X$. Let $Y$ be a subspace of $X$. If $f$ is a real linear functional on $Y$ dominated by $p$, so that*

$$f(x) \leq p(x), \qquad \text{for all } x \in Y,$$

*then there exists a linear functional $\widetilde{f}$ on $X$ that is an extension of $f$ and dominated by $p$, that is,*

$$\widetilde{f}(x) = f(x), \qquad \text{for all } x \in Y,$$

*and*

$$\widetilde{f}(x) \leq p(x), \qquad \text{for all } x \in X.$$

*Proof.* First, we employ observations made in the proof of Lemma 5.1.
We assume $Y \neq X$. For $x_0 \in X \setminus Y$, we have

$$f(x) + f(y) = f(x+y) \leq p(x+y) \leq p(x-x_0) + p(y+x_0), \quad \text{for all } x,y \in Y.$$

Hence
$$f(x) - p(x-x_0) \leq p(y+x_0) - f(y), \quad \text{for all } x,y \in Y.$$

It follows that $\sup\{f(x) - p(x-x_0) : x \in Y\} \leq \inf\{p(y+x_0) - f(y) : y \in Y\}$. Let $c = \sup\{f(x) - p(x-x_0) : x \in Y\}$. Then

$$f(x) - c \leq p(x - x_0) \quad \text{and} \quad f(y) + c \leq p(y + x_0)$$

for all $x, y \in Y$.

Let us define $\widetilde{f}(x + \lambda x_0) = f(x) + \lambda c$ for $x \in Y$. For $\lambda = 0$, we have $\widetilde{f}(x) = f(x)$, so $\widetilde{f}$ is an extension of $f$. By multiplying the inequalities displayed above by $\lambda > 0$ and substituting $x_1 = \lambda x$ and $y_1 = \lambda y$, we obtain

$$\widetilde{f}(x_1 - \lambda x_0) = f(x_1) - \lambda c \leq p(x_1 - \lambda x_0)$$

and

$$\widetilde{f}(y_1 + \lambda x_0) = f(y_1) + \lambda c \leq p(y_1 + \lambda x_0)$$

for all $x_1, y_1 \in Y$. Hence $\widetilde{f}(z) \leq p(z)$ for all $z \in \mathrm{span}(Y, \{x_0\})$.

An application of Zorn's lemma similar to that used in the proof of Theorem 5.6 completes the proof. □

In the rest of this section, we give an example of a metric vector space $X$ such that $X^* = \{0\}$. That is, the only continuous linear functional on $X$ is the zero function.

**Definition 5.1.** A function $x$ on the interval $[0, 1]$ is called a *step function* if there is a sequence of points $0 = a_0 < a_1 < \cdots < a_{n-1} < a_n = 1$ such that $x$ is constant on the intervals $[a_0, a_1), [a_1, a_2), \ldots, [a_{n-2}, a_{n-1}), [a_{n-1}, a_n]$.

**Theorem 5.9.** *The set $X_s$ of all step functions on $[0, 1]$ is a vector space over the field* **F**.

*Proof.* Clearly, it suffices to show that $X_s$ is closed under addition. Let $x$ and $y$ be step functions on $[0, 1]$ and let

$$0 = a_0 < a_1 < \cdots < a_{n-1} < a_n = 1, \quad 0 = b_0 < b_1 < \cdots < b_{m-1} < b_m = 1$$

be the corresponding sequence of points in $[0, 1]$. The set

$$\{a_0, a_1, \ldots, a_n\} \cup \{b_0, b_1, \ldots, b_m\}$$

is finite and contains the points 0 and 1. This set can be ordered as follows:

$$0 = c_0 < c_1 < \cdots < c_{k-1} < c_k = 1.$$

The function $x + y$ is constant over the intervals defined by this sequence. Hence $x + y$ is a step function. □

*Example 5.4.* For $0 < p < 1$, let $L_s^p$ be the vector space of all step functions on the interval $[0, 1]$ with the function $d$ defined by

$$d(x, y) = \int_0^1 |x(t) - y(t)|^p \, dt, \qquad \text{for all } x, y \in L_s^p.$$

The resulting space $(L_s^p, d)$ is a metric vector space (cf. Exercise 5.10).

We show that the only continuous linear functional on $L_s^p$ is the zero functional. For this we first show that the empty set and the entire space $X$ are the only convex open sets in the space $X$.

**Lemma 5.2.** *Let $Y$ be a nonempty open convex subset of the space $L_s^p$. Then $Y = L_s^p$.*

*Proof.* First, we assume that $Y$ contains the zero vector. Then there is a real number $r > 0$ such that

$$B(0, r) = \left\{ x \in L_s^p : \int_0^1 |x(t)|^p \, dt < r \right\} \subseteq Y.$$

For $n \in \mathbf{N}$ and $x \in L_s^p$, we partition $[0, 1]$ into $n$ equal subintervals by points $x_0 = 0 < x_1 < \cdots < x_n = 1$ and define

$$x^{(k)}(t) = \begin{cases} nx(t), & \text{if } t \in [x_{k-1}, x_k), \\ 0, & \text{otherwise,} \end{cases} \qquad x^{(n)}(t) = \begin{cases} nx(t), & \text{if } t \in [x_{n-1}, x_n], \\ 0, & \text{otherwise,} \end{cases}$$

for $k = 1, \ldots, n - 1$.

The function $x$ is a convex combination of these functions,

$$x(t) = \frac{1}{n} \sum_{k=1}^n x^{(k)}(t), \qquad \text{for } t \in [0, 1].$$

Let $M = \max\{|x(t)| : t \in [0, 1]\}$. We have

$$\int_0^1 |x^{(k)}(t)|^p \, dt = \int_{x_{k-1}}^{x_k} |x^{(k)}(t)|^p \, dt \leq \frac{1}{n} \cdot n^p \cdot M^p = \frac{M^p}{n^{1-p}}.$$

One can choose $n$ such that $\dfrac{M^p}{n^{1-p}} < r$ (note that $1 - p > 0$). Then

$$x^{(k)} \in B(0, r) \subseteq Y, \qquad \text{for all } 1 \leq k \leq n.$$

Since $x$ is a convex combination of the functions $x^{(k)}$ and the set $Y$ is convex, it follows that $x \in Y$. Inasmuch as $x$ is an arbitrary point in $L_s^p$, we have proved that $Y = L_s^p$.

Because every subset of $L_s^p$ is a translation of a set containing the zero vector (cf. Exercise 5.11), the only nonempty open convex subset of $L_s^p$ is $L_s^p$. □

**Theorem 5.10.** *The zero functional is the only continuous linear functional on the metric vector space $(L_s^p, d)$.*

*Proof.* Let $f$ be a continuous linear functional on $L_s^p$. For $\varepsilon > 0$, let

$$D_\varepsilon = \{\lambda \in \mathbf{F} : |\lambda| < \varepsilon\}$$

be an open disk in $\mathbf{F}$. (It is an open interval if $\mathbf{F} = \mathbf{R}$.) It is easy to see that $f^{-1}(D_\varepsilon)$ is an open convex subset of $L_s^p$. By Lemma 5.2, $f^{-1}(D_\varepsilon) = L_s^p$. It follows that $|f| < \varepsilon$ for all $\varepsilon > 0$. Hence $f$ is the zero functional on $L_s^p$.  $\square$

By the Hahn–Banach theorem, we have the following result.

**Theorem 5.11.** *The space* $(L_s^p, d)$ *is not normable.*

## 5.3 Geometric Hahn–Banach Theorems

First, we recall some notions and facts from linear algebra. In this section, the field of scalars is $\mathbf{R}$.

Let $X$ be a vector space over the field $\mathbf{R}$, $Y$ a subspace of $X$, and $a$ a vector in $X$. The set $S = a + Y$ is called an *affine set* in $X$. Thus, an affine set is a translation of a subspace (or the subspace $Y$ itself if $a \in Y$).

A subspace $Y$ of a vector space $X$ is of *codimension* 1 if $X$ is the span of $X$ and some vector $a \in X \setminus Y$. Affine sets that are translations of subspaces of codimension 1 are called *hyperplanes*. A subset $H$ of $X$ is a hyperplane if and only if $H = \{x \in X : f(x) = c, \ c \in \mathbf{R}\}$ for some nonzero linear functional $f$ (cf. Exercise 5.20).

A subset $A$ of a vector space $X$ is said to be *convex* if for every $x, y \in A$ and $\lambda \in [0, 1]$, one has $\lambda x + (1 - \lambda)y \in A$. There are many geometric assertions concerning convex sets in topological vector spaces that follow from the Hahn–Banach theorems. To give the reader a taste of these assertions, we prove two of them. For this, we need some auxiliary results.

A subset $S$ of a topological vector space $X$ is said to be *absorbing* if for every $x \in X$ there exists $r > 0$ such that $x \in \mu S$ for all $|\mu| \geq r$ (cf. Exercise 5.21).

**Lemma 5.3.** *Every neighborhood $U$ of zero in a topological vector space $X$ is absorbing.*

*Proof.* Let $x \in X$. Because $\lambda \mapsto \lambda x$ is continuous at $\lambda = 0$ (cf. Exercise 5.23), there exists $\varepsilon_x > 0$ such that $\lambda x \in U$ for all $|\lambda| < \varepsilon_x$. It follows that $x \in \mu U$, for $|\mu| \geq 2/\varepsilon_x$.  $\square$

**Lemma 5.4.** *Let $f$ be a nonzero linear functional on a topological vector space $X$. Then the image $f(V)$ of an open convex subset $V$ of $X$ is an open interval in $\mathbf{R}$.*

*Proof.* Inasmuch as $V$ convex, the set $f(V)$ is an interval in $\mathbf{R}$ (cf. Exercise 5.24).

Because $f$ is not zero, there is $x_0 \in X$ such that $f(x_0) = c \neq 0$. We set $x^* = (1/c)x_0$, so $f(x^*) = 1$. Take $y \in f(V)$. There exists $x \in V$ such that $y = f(x)$. Because scalar multiplication is continuous, there exists $\delta > 0$ such that $x + rx^* \in V$ for all $-\delta < r < \delta$. Then $y + r \in f(V)$ for all $-\delta < r < \delta$, that is, $(y - \delta, y + \delta) \subseteq f(V)$. The result follows.                    $\square$

Let $X$ be a topological real vector space and $U$ an open convex subset of $X$ containing 0. We define the *Minkowski functional* $p$ on $X$ by

$$p(x) = \inf\{t : t > 0, \ x \in tU\}.$$

Because $U$ is an absorbing set (cf. Lemma 5.3), the set $\{t : t > 0, \ x \in tU\}$ is not empty. Note that $p(0) = 0$. Therefore, $p(x) \in [0, \infty)$, $x \in X$.

**Lemma 5.5.** *The function $p$ is sublinear.*

*Proof.* (a) For $\alpha > 0$, we have

$$p(\alpha x) = \inf\{t : t > 0, \ \alpha x \in tU\} = \inf\{\alpha s : s > 0, \ x \in sU\} = \alpha p(x),$$

for every $x \in X$. Hence $p$ is positive and homogeneous.

(b) Inasmuch as $U$ is an absorbing set, for every given vectors $x$ and $y$ in $X$, there are positive numbers $s$ and $t$ such that $s^{-1}x \in U$ and $t^{-1}y \in U$. Because $U$ is a convex set, the vector

$$(s+t)^{-1}(x+y) = \frac{s}{s+t}(s^{-1}x) + \frac{t}{s+t}(t^{-1}y)$$

belongs to $U$, so $x + y \in (s+t)U$. It follows that

$$p(x + y) \leq s + t.$$

By taking the infimum over all such $s$ and $t$, we obtain

$$p(x + y) \leq p(x) + p(y),$$

so $p$ is subadditive.                                                          $\square$

**Theorem 5.12.** *Let $U$ be an open convex set in a topological vector space $X$, and $x_0$ a vector in $X$ that does not belong to $U$. Then there exists a hyperplane $H$ containing $x_0$ that does not intersect $U$.*

*Proof.* We may assume that $0 \in U$ (cf. Exercise 5.22), so $x_0 \neq 0$. Let $Y$ be the subspace $\{\lambda x_0 : \lambda \in \mathbf{R}\}$ generated by $x_0$, and $f$ the linear functional on $Y$ defined by $f(\lambda x_0) = \lambda$. Because $x_0 \notin U$, we have $p(x_0) \geq 1$. Hence for $\lambda > 0$,

$$f(\lambda x_0) = \lambda \leq \lambda p(x_0) = p(\lambda x_0).$$

The inequality $f(\lambda x_0) \leq p(\lambda x_0)$ trivially holds for $\lambda \leq 0$. Hence the functional $f$ is dominated by $p$ on $Y$. By Theorem 5.8, the functional $f$ can be extended to a linear functional $\widetilde{f}$ on $X$ dominated by $p$. The hyperplane $H$ defined by $\widetilde{f}(x) = 1$ contains $x_0$ because $\widetilde{f}(x_0) = f(x_0) = 1$. For all $x \in U$, we have

$$\widetilde{f}(x) \leq p(x) \leq 1.$$

Because $\widetilde{f}$ is a linear functional and $U$ is an open set, we must have $\widetilde{f}(x) < 1$ on $U$ (cf. Lemma 5.4). It follows that $U \cap H = \emptyset$. $\qquad\square$

*Example 5.5.* The assertion of Theorem 5.12 may not hold if $U$ is not an open set. Consider the convex set

$$U = \{(x,y) \in \ell_2^2 : -1 \leq x \leq 1, \ y \geq -\sqrt{1-x^2}\} \setminus \{(1,0)\}.$$

(This is the *epigraph* of the function $y = \sqrt{1-x^2}$ with a deleted boundary point; cf. Fig. 5.1.) There is no straight line (hyperplane) containing $(1,0)$ and not intersecting $U$.

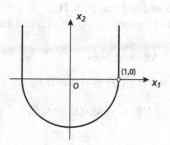

**Fig. 5.1** Example 5.5.

Let $V$ and $W$ be subsets of a topological vector space $X$. A hyperplane $H = \{x \in X : f(x) = c\}$ defined by a nonzero linear functional $f$ *separates* sets $V$ and $W$ if $f(x) \leq c$ for all $x \in V$ and $f(x) \geq c$ for all $x \in W$, or conversely, $f(x) \geq c$ for all $x \in V$ and $f(x) \leq c$ for all $x \in W$.

**Theorem 5.13.** *Let $V$ and $W$ be disjoint open convex subsets of a topological vector space $X$. Then there is a hyperplane $H$ in $X$ that separates $V$ and $W$.*

*Proof.* Choose vectors $v_0 \in V$, $w_0 \in W$, and let $x_0 = v_0 - w_0$. The set

$$U = V - W + x_0$$

is open, convex, and contains the zero vector (cf. Exercise 5.28). Because $V \cap W = \emptyset$, it follows that $x_0 \notin U$. Let $\widetilde{f}$ be as in the proof of Theorem 5.12.

Because $\widetilde{f}(x) < 1$ on $U$ and $\widetilde{f}(x_0) = 1$, we have

$$\widetilde{f}(v - w + x_0) = \widetilde{f}(v) - \widetilde{f}(w) + \widetilde{f}(x_0) < 1,$$

for every $v \in V$ and $w \in W$. Hence

$$\widetilde{f}(v) < \widetilde{f}(w), \qquad \text{for every } v \in V \text{ and } w \in W.$$

Since $V$ and $W$ are open sets, the open interval $\widetilde{f}(V)$ lies strictly to the left of the open interval $\widetilde{f}(W)$ (cf. Lemma 5.4). Therefore, there is a number $c$ such that

$$\widetilde{f}(v) < c < \widetilde{f}(w), \qquad \text{for every } v \in V \text{ and } w \in W.$$

Hence the hyperplane $H = \{x \in X : \widetilde{f}(x) = c\}$ separates sets $U$ and $W$. $\quad\square$

## 5.4 Linear Functionals on $C[a, b]$

We begin by recalling the concept of the *Riemann–Stieltjes integral* in analysis. In this section, we assume that $\mathbf{F} = \mathbf{R}$.

Let $x$ and $y$ be bounded functions on the closed interval $[a, b]$, and

$$P = \{a = t_0 < t_1 < \cdots < t_n = b\}$$

a partition of $[a, b]$. The *norm* of $P$ is defined by

$$\|P\| = \max_{1 \leq k \leq n} (t_k - t_{k-1}).$$

We arbitrarily select points $u_k \in [t_{k-1}, t_k]$, $1 \leq k \leq n$, and define a *Riemann–Stieltjes sum* by

$$S(x, y, P, U) = \sum_{k=1}^{n} x(u_k)[y(t_k) - y(t_{k-1})].$$

If there is a real number $J$ such that for each $\varepsilon > 0$, there exists $\delta > 0$ such that $|J - S(x, y, P, U)| < \varepsilon$ for every $P$ satisfying $\|P\| < \delta$, then $J$ is called the *Riemann–Stieltjes integral of $x$ with respect to $y$ on $[a, b]$*, and is denoted by

$$J = \int_a^b x(t)\, dy(t) = \int_a^b x\, dy. \tag{5.9}$$

As in the case of the Riemann integral, there are functions that are not Riemann–Stieltjes integrable. However, there is a special case of particular interest in real analysis.

**Theorem 5.14.** *If $x \in C[a,b]$ (so $x$ is a continuous function on $[a,b]$) and $w \in BV[a,b]$ (so $w$ is a function of bounded variation on $[a,b]$), then the Riemann–Stieltjes integral $\int_a^b x\,dw$ exists.*

This theorem is proved in real analysis, where the following properties of Riemann–Stieltjes integrals are established. We assume that all integrals below exist.

1. The integral in (5.9) depends linearly on functions in $C[a,b]$,

$$\int_a^b (\lambda x + \mu y)\,dw = \lambda \int_a^b x\,dw + \mu \int_a^b y\,dw,$$

so for every $w \in BV[a,b]$, formula (5.9) defines a linear functional on the space $C[a,b]$.
2. For $x \in C[a,b]$ and $w \in B[a,b]$,

$$\left| \int_a^b x\,dw \right| \le \max_{t \in [a,b]} |x(t)|\, V_a^b(w). \tag{5.10}$$

The above properties show that the Riemann–Stieltjes integral is a bounded linear functional on $C[a,b]$.

By the definition of the Riemann–Stieltjes integral, it can be approximated by Riemann–Stieltjes sums. Therefore, when $x$ is a continuous function and $w$ of bounded variation (cf. Theorem 5.14), one can use partitions $P$ of $[a,b]$ into equal subintervals and choose endpoints of the intervals in $P$ as points in the set $U$ to approximate $\int_a^b x\,dw$. Thus for $n \in \mathbf{N}$, we define the points $t_k$ of the partition $P_n$ by

$$t_k = a + \frac{b-a}{n}\,k, \qquad \text{for } 0 \le k \le n, \tag{5.11}$$

and choose $u_k = t_{k-1}$, $0 < k \le n$. The corresponding Riemann–Stieltjes sum for $x$ and $w$ depends only on $n$, so we can write

$$S_n = \sum_{k=1}^n x(t_{k-1})(w(t_k) - w(t_{k-1})).$$

By Theorem 5.14,

$$\lim_{n \to \infty} S_n = \int_a^b x\,dw.$$

We use these observations in the proof of the following theorem, which is known as Riesz's theorem for functionals on $C[a,b]$.

**Theorem 5.15.** *Every bounded linear functional $f$ on $C[a,b]$ can be represented by a Riemann–Stieltjes integral*

$$f(x) = \int_a^b x \, dw, \tag{5.12}$$

*where $w$ is of bounded variation. The norm of $f$ is the total variation of $w$:*

$$\|f\| = V_a^b(w). \tag{5.13}$$

*Proof.* Let $B[a,b]$ be the vector space of all bounded functions on $[a,b]$ endowed with the sup-norm and $f$ a bounded linear functional on $C[a,b]$. Clearly, $C[a,b]$ is a subspace of $B[a,b]$. By the Hahn–Banach theorem, $f$ has an extension $\tilde{f}$ from $C[a,b]$ to $B[a,b]$. Furthermore, $\|\tilde{f}\| = \|f\|$.

We need to define the function $w$ in (5.12). Consider the family of functions in $B[a,b]$ defined by

$$x_t(\tau) = \begin{cases} 1, & \text{if } a \leq \tau \leq t, \\ 0, & \text{if } t < \tau \leq b, \end{cases}$$

and define $w$ by

$$w(a) = 0 \qquad \text{and} \qquad w(t) = \tilde{f}(x_t), \quad \text{for } t \in (a,b].$$

We want to show that $w$ is of bounded variation. Let us define

$$\varepsilon_k = \operatorname{sgn}(w(t_k) - w(t_{k-1})), \qquad 1 \leq k \leq n,$$

where $a = t_0 < t_1 < \cdots < t_n = b$ is a partition of $[a,b]$ and sgn is the *signum function* defined by

$$\operatorname{sgn}(\tau) = \begin{cases} 0, & \text{if } \tau = 0, \\ |\tau|/\tau, & \text{if } \tau \neq 0. \end{cases}$$

Note that the signum function assumes only the values 0, 1, and $-1$, and

$$|\tau| = \operatorname{sgn}(\tau)\tau, \qquad \text{for all } \tau \in \mathbf{R}.$$

We have

$$\sum_{k=1}^{n} |w(t_k) - w(t_{k-1})| = \sum_{k=1}^{n} \varepsilon_k [w(t_k) - w(t_{k-1})]$$

$$= \varepsilon_1 \widetilde{f}(x_{t_1}) + \sum_{k=2}^{n} \varepsilon_k [\widetilde{f}(x_{t_k}) - \widetilde{f}(x_{t_{k-1}})]$$

$$= \widetilde{f}\left( \varepsilon_1 x_1 + \sum_{k=2}^{n} \varepsilon_k (x_{t_k} - x_{t_{k-1}}) \right)$$

$$\leq \|\widetilde{f}\| \left\| \varepsilon_1 x_{t_1} + \sum_{k=2}^{n} \varepsilon_k (x_{t_k} - x_{t_{k-1}}) \right\|.$$

For each $\tau \in [a,b]$, only one of the terms $x_{t_1}$ and $(x_{t_k} - x_{t_{k-1}})$, $2 \leq k \leq n$, is nonzero, and its norm is 1. Note that $\|\widetilde{f}\| = \|f\|$. Therefore,

$$\sum_{k=1}^{n} |w(t_k) - w(t_{k-1})| \leq \|f\|,$$

for every partition of $[a,b]$. Therefore, $w$ is of bounded variation. Moreover, it follows that

$$V_a^b(w) \leq \|f\|. \tag{5.14}$$

Now we prove (5.12), where $x \in C[a,b]$. For the partition $P_n$ defined by (5.11), we define a function $z_n$ by the formula

$$z_n = x(t_0) x_{t_1} + \sum_{k=2}^{n} x(t_{k-1}) [x_{t_k} - x_{t_{k-1}}].$$

Clearly, $z_n \in B[a,b]$. By the definition of $w$, we obtain

$$\widetilde{f}(z_n) = x(t_0) \widetilde{f}(x_{t_1}) + \sum_{k=2}^{n} x(t_{k-1}) [\widetilde{f}(x_{t_k}) - \widetilde{f}(x_{t_{k-1}})]$$

$$= x(t_0) w(t_1) + \sum_{k=2}^{n} x(t_{k-1}) [w(t_k) - w(t_{k-1})]$$

$$= \sum_{k=1}^{n} x(t_{k-1}) [w(t_k) - w(t_{k-1})],$$

because $w(t_0) = w(a) = 0$. The right-hand side of this chain of equalities is the Riemann–Stieltjes sum $S_n$ for the integral $\int_a^b x \, dw$. Therefore,

$$\int_a^b x \, dw = \lim_{n \to \infty} \widetilde{f}(z_n).$$

Note that

$$x_{t_k}(t) - x_{t_{k-1}}(t) = \begin{cases} 1, & \text{if } t \in (t_{k-1}, t_k], \\ 0, & \text{otherwise}, \end{cases}$$

and $z_n(a) = x(a)$. Hence $z_n(t) = x(t_{k-1})$ for $t \in (t_{k-1}, t_k]$, so

$$|z_n(t) - x(t)| = |x(t_{k-1}) - x(t)|, \qquad \text{for } t \in (t_{k-1}, t_k].$$

Inasmuch as $x$ is uniformly continuous on $[a, b]$, we have $\|z_n - x\| \to 0$ as $n \to \infty$, so $z_n \to x$ in $B[a, b]$. Because $\tilde{f}$ is a continuous functional on $B[a, b]$, and $\tilde{f}$ is an extension of $f$, we obtain

$$\int_a^b x \, dw = \lim_{n \to \infty} \tilde{f}(z_n) = \tilde{f}(x) = f(x),$$

which proves (5.12).

To prove (5.13), we evaluate the norm of $f$ from (5.12). By (5.10), we have

$$|f(x)| \le \max_{t \in [a,b]} |x(t)| \, V_a^b(w) = \|x\| \, V_a^b(w).$$

Therefore, $\|f\| \le V_a^b(w)$. Together with (5.14), this yields the desired result, $\|f\| = V_a^b(w)$. $\qquad\qquad\square$

According to the proof of Theorem 5.15, $w(a) = 0$, so $w \in NBV[a, b]$ (cf. Section 3.7). However, the space $NBV[a, b]$ is not the dual of $C[a, b]$, because $w$ in Theorem 5.15 is not unique.

*Example 5.6.* The zero function and the function $w$ such that $w(b) = 1$, and $w(t) = 0$ otherwise, are of bounded variation, and both define the zero functional on $C[a, b]$.

## 5.5 Reflexivity

We begin with a simple example motivating our presentation in this section.

*Example 5.7.* It is known from linear algebra that every linear functional on the vector space $\ell_2^n$ is uniquely defined by a vector $\lambda = (\lambda_1, \ldots, \lambda_n) \in \mathbf{F}^n$ and is of the form

$$f_\lambda(x) = \lambda_1 x_1 + \cdots + \lambda_n x_n, \qquad \text{for } x = (x_1, \ldots, x_n) \in \ell_2^n. \qquad (5.15)$$

By the Cauchy–Schwarz inequality, we have

$$|f_\lambda(x)| = \left| \sum_{k=1}^n \lambda_k x_k \right| \le \sum_{k=1}^n |\lambda_k x_k| \le \sqrt{\sum_{k=1}^n |\lambda_k|^2} \sqrt{\sum_{k=1}^n |x_k|^2},$$

that is, $|f_\lambda(x)| \le \|\lambda\|_2 \|x\|_2$. It follows that $\|f_\lambda\| \le \|\lambda\|_2$. By substituting $x = \overline{\lambda}$ into (5.15), we obtain

$$|f_\lambda(\overline{\lambda})| = \sum_{k=1}^{n} |\lambda_k|^2 = \|\lambda\|_2^2,$$

which implies $\|f_\lambda\| \ge \|\lambda\|_2$. It follows that $\|f_\lambda\| = \|\lambda\|_2$. Therefore, the dual $(\ell_2^n)^*$ of the space $\ell_2^n$ is isomorphic to the space $\ell_2^n$ itself (cf. Corollary 5.1).

One can consider the right-hand side of the Eq. (5.15) a function of $\lambda \in (\ell_2^n)^*$ for a given $x \in \ell_2^n$:

$$g_x(\lambda) = \lambda_1 x_1 + \cdots + \lambda_n x_n.$$

It is clear that $g_x$ is a linear functional on $(\ell_2^n)^*$. By repeating the argument used for $f_\lambda(x)$, it is easy to show that $\|g_x\| = \|x\|_2$. It follows that the mapping $x \mapsto g_x$ (which is clearly onto) is an isomorphism between spaces $\ell_2^n$ and $((\ell_2^n)^*)^*$.

The dual space $X^*$ of a normed space $X$ is a normed space itself. Thus we may consider the dual space of the space $X^*$. This space is called the *second dual space* of $X$ and is denoted by $X^{**}$ (so $X^{**} = (X^*)^*$). The usual interpretation of the expression $f(x)$ is that $f$ is a given function and $x$ is a variable. However, we can interpret $f(x)$ in a "dual" way by fixing $x$ and considering $f$ a variable (cf. Example 5.7). This interpretation leads to the following definition:

**Definition 5.2.** The mapping $J : X \to X^{**}$ defined by $x \mapsto g_x$, where

$$g_x(f) = f(x), \qquad f \in X^*,$$

is called the *canonical mapping* of $X$ into $X^{**}$.

To justify this definition, we note that for every $x \in X$, the function $g_x$ is a linear functional on $X^*$, because

$$g_{\lambda x + \mu y}(f) = f(\lambda x + \mu y) = \lambda f(x) + \mu f(y) = \lambda g_x(f) + \mu g_y(f),$$

for all $\lambda, \mu \in \mathbf{F}$. Furthermore, by Exercise 5.9,

$$\|g_x\| = \sup_{\substack{f \in X^* \\ f \ne 0}} \frac{|g_x(f)|}{\|f\|} = \sup_{\substack{f \in X^* \\ f \ne 0}} \frac{|f(x)|}{\|f\|} = \|x\|,$$

so $g_x$ is bounded.

It follows from the last displayed formula that the canonical mapping $J$ is an isomorphism of the normed space $X$ onto its image $J(X)$ in the second dual space $X^{**}$. In general, $J(X)$ is a proper subspace of $X^{**}$. The case $J(X) = X^{**}$ is of particular importance in functional analysis.

**Definition 5.3.** A normed space is said to be *reflexive* if

$$J(X) = X^{**},$$

where $J$ is the canonical mapping.

Note that a reflexive space is a Banach space, because the dual space of every space is a Banach space (cf. Theorem 4.8).

A prototypical example of a reflexive space is the sequence space $\ell_p$ with $p \in (1, \infty)$. By Theorem 5.4, the dual $\ell_p^*$ of $\ell_p$ is isomorphic to the space $\ell_q$, where $q$ and $p$ are conjugate exponents. By exchanging the roles of $p$ and $q$, we conclude (by the same theorem) that the second dual $\ell_p^{**}$ is isomorphic to the space $\ell_p$ itself. However, this argument shows only that the spaces $\ell_p^{**}$ and $\ell_p$ are isomorphic but does not prove that the canonical mapping is the required isomorphism. For this, we prove the following theorem.

**Theorem 5.16.** *The space $\ell_p$ is reflexive.*

*Proof.* Because the canonical mapping $J$ is an isomorphism of the space $X$ onto its image $J(X)$, it suffices to show that $J$ is surjective. Thus we need to show that for every $h \in \ell_p^{**}$ there exists $x = (x_n) \in \ell_p$ such that $h = g_x$. We may consider $h$ a bounded linear functional on the space $\ell_q$, which is isomorphic to $\ell_p^*$. By the proof of Theorem 5.4 (exchanging $p$ and $q$ and using (5.5)), we have

$$h(f) = \sum_{k=1}^{\infty} x_k f_k, \qquad \text{for } x = (x_n) \in \ell_p \text{ and } f = (f_n) \in \ell_q.$$

By the same theorem (cf. (5.5)),

$$\sum_{k=1}^{\infty} x_k f_k = f(x), \qquad \text{for } x = (x_n) \in \ell_p \text{ and } f = (f_n) \in \ell_q.$$

Inasmuch as $g_x(f) = f(x)$, we have $h(f) = g_x(f)$ for all $f \in \ell_p^*$. Hence $h = g_x$, and the result follows. $\square$

It is not difficult to prove that the spaces $c_0$ and $c$ are not reflexive. By Theorems 5.2 and 5.3, $c_0^* = c^* = \ell_1$, and by Theorem 5.1, $\ell_1^* = \ell_\infty$. Therefore,

$$c_0^{**} = c^{**} = \ell_\infty.$$

The space $c_0$ (and $c$) is separable (cf. Exercise 3.18), whereas the space $\ell_\infty$ is not (cf. Theorem 3.7). Hence $c_0^{**}$ and $\ell_\infty$ are not isomorphic. It follows that $c_0$ (and $c$) is not reflexive.

In the conclusion of this section, we establish some results concerning the relation between the separability and reflexivity properties of a normed space. First, we prove a lemma.

**Lemma 5.6.** *Let $Y$ be a closed proper subspace of a normed space $X$. Then there exists $f \in X^*$ such that*

$$\|f\| = 1 \qquad and \qquad f(x) = 0, \quad for \; x \in Y.$$

*Proof.* For $x_0 \in X \setminus Y$, we define $Z = \text{span}\{Y, x_0\}$ and

$$g(x + \lambda x_0) = \lambda \qquad \text{for all } x \in Y \text{ and } \lambda \in \mathbf{F}.$$

The linear functional $g$ is bounded on $Z$. Indeed, because $X \setminus Y$ is an open set, there exists $r > 0$ such that $\|y - x_0\| \geq r$ for all $y \in Y$. Therefore,

$$\|x + \lambda x_0\| = |\lambda| \|(-1/\lambda)x - x_0\| \geq |\lambda| r \qquad \text{on } Z.$$

From this we have

$$|g(x + \lambda x_0)| = |\lambda| \leq \frac{1}{r}\|x + \lambda x_0\| \qquad \text{on } Z.$$

Hence $\|g\| \leq 1/r$, so $g$ is a bounded linear functional on $Z$. Clearly, $g$ vanishes on $Y$. By the Hahn–Banach theorem, $g$ has an extension to a bounded linear functional $\widetilde{g}$ on $X$ such that $\widetilde{g}(x) = 0$ on $Y$. Because $\widetilde{g}(x_0) = 1$, the norm $\|\widetilde{g}\|$ is nonzero. We obtain the desired result by setting $f = (1/\|\widetilde{g}\|)\widetilde{g}$. $\qquad \square$

**Theorem 5.17.** *If the dual space $X^*$ of a normed space $X$ is separable, then $X$ is also separable.*

*Proof.* Because $X^*$ is separable, the unit sphere $S = \{f \in X^* : \|f\| = 1\}$ is separable (cf. Exercise 2.26). Let $\{f_n\}_{n \in \mathbf{N}}$ be a countable dense subset of $S$. Because $f_n \in S$, we have $\sup\{|f_n(x)| : \|x\| = 1\} = 1$. Therefore, for each $n$, we can find $x_n \in X$ such that

$$\|x_n\| = 1 \qquad and \qquad |f_n(x_n)| \geq \frac{1}{2}.$$

Let $Y$ be the closure of $\text{span}\{x_n\}_{n \in \mathbf{N}}$. The set of all linear combinations of the $x_n$ with rational coefficients is a countable dense subset of $Y$. So it suffices to show that $Y = X$. Indeed, otherwise, by Lemma 5.6, there exists $f \in X^*$ such that

$$\|f\| = 1 \qquad and \qquad f(x) = 0 \quad \text{for all } x \in Y.$$

Clearly, $f(x_n) = 0$ for all $n \in \mathbf{N}$. Furthermore,

$$\frac{1}{2} \leq |f_n(x_n)| = |f_n(x_n) - f(x_n)| = |(f_n - f)(x_n)|$$
$$\leq \|f_n - f\| \|x_n\| = \|f_n - f\|,$$

contradicting our assumption that $\{f_n\}_{n\in\mathbf{N}}$ is a dense subset of $S$. Thus $X$ is separable.                                                                            □

Suppose that $X$ is a reflexive and separable space. Because the second dual $X^{**} = (X^*)^*$ is isomorphic to $X$, it is also separable. Then by Theorem 5.17, the space $X^*$ is separable. From this it follows that a separable normed space with an inseparable dual cannot be reflexive. In particular, the separable space $\ell_1$ is not reflexive, because its dual, $\ell_\infty$, is not separable.

To conclude this section, we show that the space $C[0,1]$ with the sup-norm is not reflexive.

First, we introduce a "triangular" function on $[0,1]$. For $0 < \alpha < \beta < 1$, we define a piecewise linear function $\triangle_{\{\alpha,\beta\}}$ that has nodes at points $(0,0)$, $(\alpha,0)$, $((\alpha+\beta)/2,1)$, $(\beta,0)$, and $(1,0)$ (cf. Fig. 5.2).

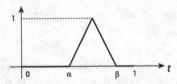

**Fig. 5.2** Graph of the function $\triangle_{\{\alpha,\beta\}}$.

Consider the decreasing sequence

$$1 > \beta_1 > \alpha_1 > \beta_2 > \alpha_2 > \cdots > \beta_n > \alpha_n > \cdots$$

converging to zero. For each $n \in \mathbf{N}$, we define $f_n(t) = \triangle_{\{\alpha_n,\beta_n\}}(t)$ on $[0,1]$.

Let $x = (x_n) \in c_0$, so $x_n \to 0$. It is not difficult to show that the function $f_x(t) = \sum_{k=1}^{\infty} x_k f_k(t)$ is a continuous function on $[0,1]$ and that $\|f_x\| = \|x\|$. Hence the correspondence $T : x \mapsto f_x$ is an isomorphism of the normed space $c_0$ into the normed space $C[0,1]$. Because $c_0$ is a complete space, the image $T(c_0)$ is a closed subspace of $C[0,1]$ (cf. Theorem 4.11), which is not reflexive, because $c_0$ is not reflexive. By Exercise 5.32, $C[0,1]$ is not a reflexive space.

## Notes

On a finite-dimensional normed space, every linear functional is bounded (cf. Exercise 5.1).

The Hahn–Banach theorems have many applications. One of them is Riesz's theorem (Theorem 5.15 in Section 5.4), providing the representation of bounded linear functionals on the normed space $C[a,b]$ by the Riemann–Stieltjes integral.

Corollary 5.2 guarantees the existence of sufficiently many continuous linear functionals on a seminormed vector space with a nontrivial seminorm.

Example 5.4 shows that there are topological vector spaces with no non-trivial continuous linear functionals on them. It is not difficult to show that the metric space $(X, d)$ in this example is not complete. This can be remedied as follows. We omit the details.

For $0 < p < 1$, let $L^p$ be the vector space of all Lebesgue measurable functions $f$ on $[0, 1]$ for which

$$\int_0^1 |f|^p < \infty,$$

with the usual identification of functions that coincide almost everywhere. The space $L^p$ can be made into a complete metric vector space by defining

$$d(x, y) = \int_0^1 |x - y|^p, \qquad \text{for all } x, y \in L^p.$$

It can be shown that the metric space $(X, d)$ in Example 5.4 is dense in $(L^p, d)$.

**Theorem.** (Rudin 1973 p. 36) The zero functional is the only continuous linear functional on the metric vector space $(L^p, d)$ (cf. Exercise 5.12).

If a normed space $X$ is reflexive, it is isomorphic to $X^{**}$ under the canonical mapping. The converse does not hold in general. In 1951, Robert C. James published a paper entitled "A non-reflexive Banach space isometric with its second conjugate space" in the *Proceedings of the National Academy of Sciences, USA*, in which a counterexample was constructed.

# Exercises

**5.1.** Prove that every linear functional on a finite-dimensional normed space is bounded. (Hint: Use Lemma 4.1.)

**5.2.** Let $x = \sum_{k=1}^\infty x_k e_k$ be the expansion of a vector $x$ with respect to a Schauder basis $(e_n)$ in a normed space $X$. Show that for every $k \in \mathbf{N}$, the mapping $x \mapsto x_k$ is a linear functional on $X$. (These functionals are called *coordinate* or *biorthogonal functionals*. They are all bounded (cf. Theorem 6.4)).

**5.3.** Prove Theorem 5.3.

**5.4.** Show that every sequence $(b_n) \in c_{00}$ defines a continuous linear functional on the space $s$ by (5.8).

**5.5.** In the proof of Lemma 5.1, show that one can choose

$$c = \sup_{y' \in Y} \left[ -p(x_0 - y') + f(y') \right]$$

or

$$c = \inf_{y'' \in Y} [p(x_0 - y'') + f(y'')].$$

**5.6.** Prove Corollary 5.3.

**5.7.** Give an example of a normed space and a proper subspace of that space for which the assertion of Corollary 5.4 does not hold.

**5.8.** Show that $\beta = 0$ in Example 5.1.

**5.9.** Prove that for every $x$ in a normed space $X$, the following identity holds:

$$\|x\| = \sup \left\{ \frac{|f(x)|}{\|f\|} : f \in X^* \ f \neq 0 \right\}.$$

**5.10.** For $0 < p < 1$, let $X$ be the vector space of all step functions on the interval $[0, 1]$ with the function $d$ defined by

$$d(x, y) = \int_0^1 |x(t) - y(t)|^p \, dt, \qquad \text{for all } x, y \in X.$$

Show that $(X, d)$ is a metric vector space.

**5.11.** Let $Y$ be a convex subset of a vector space $X$. Show that for every $x \in X$, the set $Y + x = \{y + x : y \in Y\}$ is convex.

**5.12.** Let $(b_n)$ be a bounded sequence in $\mathbf{F}$ and $(x_k) \in \ell_p$, where $p \in (0, 1)$. Show that

$$f(x) = \sum_{k=1}^{\infty} b_k x_k$$

is a continuous linear functional on the metric vector space $(\ell_p, d)$ (cf. Section 3.3).

**5.13.** Show that $p(x) = \limsup x_n$, where $x = (x_n) \in \ell_\infty$, $x_n \in \mathbf{R}$, defines a sublinear functional on $\ell_\infty$.

**5.14.** Show that a sublinear functional $p$ on a real vector space $X$ satisfies $p(0) = 0$ and $p(-x) \geq -p(x)$, for every $x \in X$.

**5.15.** Let $p$ be a subadditive functional on a normed space $X$. Show that continuity of $p$ at $x = 0$ implies continuity of $p$ on $X$.

**5.16.** Prove that a linear combination of sublinear functionals is a sublinear functional.

**5.17.** Show that $x = y$ if $f(x) = f(y)$ for every bounded linear functional on a normed space $X$.

**5.18.** Prove that a subset $S$ of a vector space is affine if and only if it is closed under linear combinations

$$\lambda_1 x_1 + \lambda_2 x_2 + \cdots + \lambda_k x_k \in S$$

for all vectors $x_1, \ldots, x_k \in S$ and scalars $\lambda_1, \ldots, \lambda_k \in \mathbf{F}$ with $\sum_{i=1}^{k} \lambda_i = 0$.

**5.19.** Prove that a nonempty subset $S$ of a vector space $X$ is an affine set if and only if it is the inverse image of a singleton under a linear map of $X$ into some vector space $Y$.

**5.20.** Show that a subset $S$ of a vector space $X$ is a hyperplane if and only if there exist a linear functional $f : X \to \mathbf{F}$ and $c \in \mathbf{F}$ such that

$$S = \{x \in X : f(x) = c\}.$$

In particular, the subspaces of codimension 1 of a vector space $X$ are exactly the null spaces of linear functionals on $X$.

**5.21.** Give an example of a nonabsorbing set in a topological vector space.

**5.22.** Show that $0 \in U$ is a valid assumption in the proof of Theorem 5.12.

**5.23.** Suppose $X$ is a topological vector space. Prove that for every $x_0 \in X$, the mapping $\mathbf{F} \to X$ defined by $\lambda \mapsto \lambda x_0$ is continuous.

**5.24.** (a) Prove that the image of a convex set under linear mappings of vector spaces is convex.
(b) Prove that intervals are the only convex subsets of $\mathbf{R}$.

**5.25.** (a) Let $p$ be the Minkowski functional for a convex set $U \subseteq X$. Show that for $x \neq 0$, $p(x) = 0$ if and only if $x \in tU$ for every $t > 0$, so $U$ is "unbounded in the direction of the vector $x$."
(b) Show that $p(x) \leq 1$ if $x \in U$, and $p(x) \geq 1$ if $x \notin U$.

**5.26.** If $f$ is a real linear functional on a real vector space $X$, then the image of a convex subset of $X$ is an interval in $\mathbf{R}$.

**5.27.** If $V$ and $W$ are open convex subsets of a vector space, then so is their difference
$$V - W = \{v - w : v \in V, \, w \in W\}.$$

**5.28.** Let $V$ and $W$ be open convex subsets of a topological vector space $X$. For $v_0 \in V$ and $w_0 \in W$, define $x_0 = v_0 - w_0$ and show that the set

$$U = V - W + x_0$$

is open and convex, and that $x_0 \in U$.

**5.29.** Let $M$ be the set of sequences in the real space $c_{00}$ for which the leading nonzero term is positive. Show that the sets $M$ and $-M$ are convex and disjoint, but they cannot be separated by a hyperplane (cf. Theorem 5.13). (Recall that $c_{00}^* = \ell_1$.)

**5.30.** Let $C[0,1]$ be the vector space of continuous functions on $[0,1]$. Define $\delta(x) = x(0)$ for $x \in C[0,1]$.

(a) Show that $\delta$ is a bounded linear functional if $C[0,1]$ is endowed with the sup-norm. Find the norm of $\delta$.
(b) Show that $\delta$ is an unbounded linear functional if $C[0,1]$ is endowed with the norm
$$\|x\| = \int_0^1 |x(t)|\, dt.$$

**5.31.** Show that the dual of a reflexive normed space is also reflexive.

**5.32.** Prove that a closed subspace of a reflexive Banach space is reflexive.

**5.33.** Prove that a Banach space is reflexive if and only if its dual space is reflexive.

**5.34.** Show that if a normed space has $n$ linearly independent vectors, then so does its dual space.

# Chapter 6
# Fundamental Theorems

In this chapter we present three theorems that are known as the "fundamental theorems" of functional analysis (together with the Hahn–Banach theorems presented in Chapter 5). These are the uniform boundedness theorem (cf. Section 6.1), the open mapping theorem (cf. Section 6.2), and the closed graph theorem (cf. Section 6.3). These theorems are the cornerstones of the theory of Banach spaces and are ubiquitous in functional analysis.

The results of the fundamental theorems illustrate the power of the concept of a Banach space as opposed to incomplete normed spaces. In Section 6.4, we give several examples and counterexamples, showing that the results of the first three sections do not hold in some incomplete normed spaces.

## 6.1 Uniform Boundedness Theorem

**Theorem 6.1. Uniform boundedness theorem.** *If $\mathcal{F} = \{T_j\}_{j \in J}$ is a family of bounded operators from a Banach space $X$ to a normed space $Y$ such that $\sup_{j \in J} \|T_j x\| < \infty$ for every $x \in X$, then $\sup_{j \in J} \|T_j\| < \infty$.*

In words: If a family of bounded operators from a Banach space to a normed space is pointwise bounded, then it is norm bounded.

There are several proofs of this important theorem. A proof utilizing the Baire category theorem (cf. Section 2.6) is presented below.

*Proof.* For $j \in J$ and $n \in \mathbf{N}$, we define

$$A_{j,n} = \{x \in X : \|T_j x\| \leq n\}.$$

Because bounded operators are continuous functions and the norm is also a continuous function, each set $A_{j,n}$ is a closed subset of $X$, hence so are the sets

$$A_n = \bigcap_{j \in J} A_{j,n}, \qquad \text{for } n \in \mathbf{N}.$$

© Springer International Publishing AG, part of Springer Nature 2018
S. Ovchinnikov, *Functional Analysis*, Universitext,
https://doi.org/10.1007/978-3-319-91512-8_6

Inasmuch as the family $\{T_j\}_{j\in J}$ is pointwise bounded, we have

$$X = \bigcup_{n=1}^{\infty} A_n.$$

By the Baire category theorem (note that $X$ is a complete metric space), there is a natural number $m$ such that the set $A_m$ contains an open ball, say

$$B(x_0, r) \subseteq A_m, \qquad \text{where } x_0 \in A_m \text{ and } r > 0.$$

It is clear that for every vector $y \in B(x_0, r)$, we have $\|T_j y\| \le m$ for all $j \in J$.

To show that the family $\{T_j\}_{j\in J}$ is norm bounded, it suffices to find a constant $C$ such that for all $x \in X$ with $\|x\| = 1$ and all $j \in J$, $\|T_j x\| \le C$. For this we use the "scale and translate" technique. We scale the vector $x$ by the factor $r/2$ and translate the resulting vector by $x_0$ to obtain the vector

$$y = \frac{r}{2}x + x_0,$$

which clearly belongs to the ball $B(x_0, r)$. Hence $\|T_j y\| \le m$ for all $j \in J$. Finally, we have

$$\|T_j x\| = \left\|\frac{2}{r}T_j(y - x_0)\right\| \le \frac{2}{r}\|T_j y\| + \frac{2}{r}\|T_j x_0\| \le \frac{4m}{r}, \qquad \text{for all } j \in J.$$

Thus we can choose $C = (4m)/r$. $\qquad\qquad\square$

There are many applications of the uniform boundedness theorem. Below we present two examples.

*Example 6.1.* Let $p$ and $q$ be conjugate exponents (cf. Section 1.2). Suppose that $(a_n)$ is a sequence of complex numbers such that $\sum_{k=1}^{\infty} a_k x_k < \infty$ for every $(x_k) \in l^p$. We show that then $\sum_{k=1}^{\infty} |a_k|^q < \infty$, so $(a_k) \in \ell^q$.

We may assume that $(a_n)$ has no zero terms (cf. Exercise 6.1). Let

$$f_n(x) = \sum_{k=1}^{n} a_k x_k.$$

Clearly, each $f_n$ is a linear functional on $\ell^p$. For a given $x = (x_n) \in \ell^p$,

$$\sup_{n\in\mathbf{N}} \|f_n(x)\| = \sup_{n\in\mathbf{N}} \left|\sum_{k=1}^{n} a_k x_k\right| < \infty,$$

because $\sum_{k=1}^{\infty} a_k x_k < \infty$. (A convergent sequence is bounded.) Therefore, the family $\{f_n\}_{n\in\mathbf{N}}$ is pointwise bounded.

By Hölder's inequality, we have

$$|f_n(x)| = \Big| \sum_{k=1}^{n} a_k x_k \Big| \le \sum_{k=1}^{n} |a_k x_k| \le \Big( \sum_{k=1}^{n} |a_k|^q \Big)^{1/q} \Big( \sum_{k=1}^{n} |x_k|^p \Big)^{1/p}$$

$$\le \Big( \sum_{k=1}^{n} |a_k|^q \Big)^{1/q} \|x\|.$$

Hence $\|f_n\| \le \Big( \sum_{k=1}^{n} |a_k|^q \Big)^{1/q}$.

On the other hand, for

$$x_k = \frac{\bar{a}_k |a_k|^{q-2}}{\Big( \sum_{k=1}^{n} |a_k|^q \Big)^{1-1/q}},$$

where $1 \le k \le n$ and $x_k = 0$ for $k > n$, we have

$$|f_n(x)| = \frac{\sum_{k=1}^{n} |a_k|^q}{\Big( \sum_{k=1}^{n} |a_k|^q \Big)^{1-1/q}} = \Big( \sum_{k=1}^{n} |a_k|^q \Big)^{1/q}.$$

Hence $\|f_n\| = \Big( \sum_{k=1}^{n} |a_k|^q \Big)^{1/q}$ for $n \in \mathbf{N}$. By the uniform boundedness theorem, the series $\sum_{k=1}^{\infty} |a_k|^q$ converges.

*Example 6.2.* Suppose that $(a_n)$ is a sequence of complex numbers such that $\sum_{k=1}^{\infty} |a_k x_k| < \infty$ for every $(x_k) \in c_0$. We show that this implies

$$\sum_{k=1}^{\infty} |a_k| < \infty.$$

We may assume that $(a_n)$ has no zero terms (cf. Exercise 6.1). Let

$$f_n(x) = \sum_{k=1}^{n} a_k x_k.$$

Clearly, each $f_n$ is a linear functional on $c_0$. For a given $x = (x_n) \in c_0$,

$$\sup_{n \in \mathbf{N}} \|f_n(x)\| = \sup_{n \in \mathbf{N}} \Big| \sum_{k=1}^{n} a_k x_k \Big| \le \sup_{n \in \mathbf{N}} \sum_{k=1}^{n} |a_k x_k| < \infty,$$

because $\sum_{k=1}^{\infty} |a_k x_k| < \infty$. (A convergent sequence is bounded.) Therefore, the family $\{f_n\}_{n \in \mathbf{N}}$ is pointwise bounded.

We have

$$\|f_n\| = \sup_{\|x\|=1} \Big| \sum_{k=1}^{n} a_k x_k \Big| \le \sup_{\|x\|=1} \sum_{k=1}^{n} |a_k||x_k| \le \sum_{k=1}^{n} |a_k|.$$

On the other hand, for the sequence $x = (x_n) \in c_0$ defined by

$$x_k = \begin{cases} |a_k|/a_k, & \text{for } 1 \le k \le n, \\ 0, & \text{for } k > n, \end{cases}$$

we have $\|x\| = 1$ and $\|f_n(x)\| = \sum_{k=1}^{n} |a_k|$. Hence $\|f_n\| = \sum_{k=1}^{n} |a_k|$. By the uniform boundedness theorem, $\sum_{k=1}^{\infty} |a_k| < \infty$.

## 6.2 Open Mapping Theorem

We recall some definitions from linear algebra. If $X$ is a vector space over a field $\mathbf{F}$ and $A, B$ are subsets of $X$, then:

1. For $k \in \mathbf{F}$, $kA = \{kx : x \in X\}$.
2. $A + B = \{a + b : a \in A, \ b \in B\}$.
3. $A - B = A + (-1)B = \{a - b : a \in A, \ b \in B\}$.
4. For $y \in X$, $A + y = A + \{y\} = \{x + y : x \in X\}$.
5. A subset $A \subseteq X$ is said to be symmetric if $x \in A$ implies $-x \in A$ for all $x \in A$.

The following theorem establishes properties of the topological closure with respect to algebraic operations on subsets of a topological vector space.

**Theorem 6.2.** *Let $A$ and $B$ be subsets of a topological vector space. Then:*

(a) $k\overline{A} = \overline{kA}, \quad k \in \mathbf{F}$.
(b) $\overline{A} + \overline{B} \subseteq \overline{A + B}$.

*Proof.* (a) Let $a \in \overline{A}$ and let $U$ be a neighborhood of $ka$. Because multiplication by a scalar is a continuous operation, there is a neighborhood $U_1$ of $a$ such that $kU_1 \subseteq U$. Because $a \in \overline{A}$, there exists $x \in A \cap U_1$. Then $kx \in kA \cap kU_1 \subseteq kA \cap U$. Hence $kA \cap U \ne \varnothing$, implying $ka \in \overline{kA}$. We have proved that $k\overline{A} \subseteq \overline{kA}$. For the opposite inclusion, we may assume $k \ne 0$ and apply the previous argument as follows:

$$\frac{1}{k}\overline{kA} \subseteq \overline{\frac{1}{k}kA} = \overline{A},$$

so $\overline{kA} \subseteq k\overline{A}$. It follows that $k\overline{A} = \overline{kA}$.

(b) Let $a \in \overline{A}$, $b \in \overline{B}$, and let $U$ be a neighborhood of $a + b$. We need to show that the intersection $(A + B) \cap U$ is not empty. Inasmuch as vector addition is a continuous operation, there are neighborhoods $U_1$ and $U_2$ of vectors $a$ and $b$, respectively, such that $U_1 + U_2 \subseteq U$. Because $a \in \overline{A}$ and $b \in \overline{B}$, there are vectors $x \in U_1 \cap A$ and $y \in U_2 \cap B$. Then $x + y \in (A + B) \cap U$. Hence, $(A + B) \cap U \ne \varnothing$. $\qquad \square$

The proof of the following corollary is left as an exercise (cf. Exercise 6.8).

**Corollary 6.1.** $\overline{A} - \overline{B} \subseteq \overline{A - B}$.

**Definition 6.1.** A mapping $f : X \to Y$ of a topological space $X$ into a topological space $Y$ is said to be *open* if the image of every open set in $X$ is open in $Y$.

The goal of this section is to establish the following result.

**Theorem 6.3. Open mapping theorem.** *Let $T$ be a bounded operator from a Banach space $X$ onto a Banach space $Y$. Then $T$ is an open mapping. Accordingly, if $T$ is a bijection, then it is a homeomorphism, that is, $T^{-1}$ is continuous.*

Note that every open linear mapping $T$ of a normed space $X$ to a normed space $Y$ is necessarily onto. Indeed, the image $T(X)$ is an open subspace of $Y$ and therefore contains a ball $B(0, r)$ for some $r > 0$. Every vector $y \in Y$ belongs to the ball $B(0, 2\|y\|) = (2\|y\|/r)B(0, r) \subseteq T(X)$. Hence $T$ is onto.

Before proving Theorem 6.3, we prove two lemmas.

**Lemma 6.1.** *Let $U$ be a convex symmetric neighborhood of the origin in a topological vector space $X$, and $T$ a linear operator from $X$ to a topological vector space. Then*

$$\overline{T(U)} - \overline{T(U)} \subseteq 2\overline{T(U)}.$$

*Proof.* Inasmuch as $U$ is convex and symmetric, we have

$$\frac{x - y}{2} = \frac{1}{2}x + \frac{1}{2}(-y) \in U,$$

for every $x, y \in U$. Hence $T(U) - T(U) \subseteq 2T(U)$. Therefore, by Corollary 6.1 and Theorem 6.2 (a),

$$\overline{T(U)} - \overline{T(U)} \subseteq \overline{T(U) - T(U)} \subseteq 2\overline{T(U)},$$

which is the desired result.                                                                     □

Note that open balls $B(0, r)$ in a normed space are convex (cf. Exercise 6.9 (a)) and obviously symmetric. In what follows, we use the notation $B_r = B(0, r)$.

**Lemma 6.2.** *Let $T$ be a bounded operator from a Banach space $X$ onto a Banach space $Y$. Then the image $T(B_r)$ of a ball in $X$ contains a ball about the origin in $Y$.*

*Proof.* Because

$$X = \bigcup_{k=1}^{\infty} kB_{r/2}$$

and $T$ is onto, we have

$$Y = \bigcup_{k=1}^{\infty} kT(B_{r/2}) = \bigcup_{k=1}^{\infty} k\overline{T(B_{r/2})}.$$

Note that by taking closures we do not add new points, because the union is the entire space $Y$. Our choice of $r/2$ over $r$ will become evident in the next paragraph.

By the Baire category theorem ($Y$ is a complete space), for some $k \in \mathbf{N}$, we have $k\overline{T(B_{r/2})}$, and hence $\overline{T(B_{r/2})}$ contains some ball $B(x_0, \rho)$. Then $\overline{T(B_{r/2})} - x_0$ contains the ball $B_\rho$. By Lemma 6.1,

$$\overline{T(B_{r/2})} - x_0 \subseteq \overline{T(B_{r/2})} - \overline{T(B_{r/2})} \subseteq 2\overline{T(B_{r/2})} = \overline{T(B_r)}.$$

It follows that $\overline{T(B_r)}$ contains the open ball $B_\rho$. Hence $\overline{T(B_{r/2^n})}$ contains the open ball $B_{\rho/2^n}$ for $n = 1, 2, \ldots$. We use this observation in the rest of the proof.

Although we cannot claim that $T(B_r)$ contains the ball $B_\rho$, we show below that it contains the ball $B_{\rho/2}$, therefore establishing the desired result.

Let $y$ be an arbitrary element of $B_{\rho/2}$. Because $B_{\rho/2} \subseteq \overline{T(B_{r/2})}$, there exists $x_1 \in B_{r/2}$ such that

$$\|y - Tx_1\| < \frac{\rho}{4}.$$

Hence the vector $y - Tx_1$ belongs to $B_{\rho/4}$. Because $B_{\rho/4} \subseteq \overline{T(B_{r/4})}$, there exists $x_2 \in B_{r/4}$ such that

$$\|(y - Tx_1) - Tx_2\| < \frac{\rho}{8}.$$

By repeating this process, we construct a sequence of vectors $(x_n)$ such that

$$\|x_n\| < \frac{r}{2^n} \quad \text{and} \quad \left\| y - \sum_{k=1}^{n} Tx_n \right\| < \frac{\rho}{2^{n+1}}, \tag{6.1}$$

for $n \in \mathbf{N}$. By the first property in (6.1), the sequence of sums $(\sum_{k=1}^{n} x_k)$ is Cauchy. Indeed, we have for $m < n$,

$$\left\| \sum_{k=1}^{n} x_k - \sum_{k=1}^{m} x_k \right\| = \left\| \sum_{k=m+1}^{n} x_k \right\| \le \sum_{k=m+1}^{n} \|x_k\| < \sum_{k=m+1}^{n} \frac{r}{2^k} = \frac{r}{2^m} - \frac{r}{2^n},$$

which can be made less than every prescribed positive number by choosing sufficiently large $m$ and $n$. Because the space $X$ is a Banach space, the sequence $(\sum_{k=1}^{n} x_k)$ converges to some $x \in X$. By the continuity of the norm (cf. Exercise 3.6), we have

$$\|x\| = \left\| \sum_{k=1}^{\infty} x_k \right\| \le \sum_{k=1}^{\infty} \|x_k\| < \sum_{k=1}^{\infty} \frac{r}{2^k} = r.$$

Hence $x \in B_r$. By the second property in (6.1), the sequence $(\sum_{k=1}^{n} Tx_k)$ converges to $y$. Inasmuch as $T$ is continuous, we have $y = Tx$. Therefore, $y \in T(B_r)$. It follows that $B_{\rho/2} \subseteq T(B_r)$, as desired. $\qquad\square$

Now we proceed with the proof of the open mapping theorem.

*Proof.* Let $U$ be a nonempty open set in $X$, and $y$ a vector in $T(U)$. There exist $x \in U$ such that $y = Tx$ and a ball $B(x, r)$ that is contained in $U$. By Lemma 6.2, the image $T(B(x, r) - x)$ of the ball $B_r = B(x, r) - x$ must contain a ball $B_s$ centered at the origin of $Y$. Because

$$T(B(x,r) - x) = T(B(x,r)) - Tx = T(B(x,r)) - y \subseteq T(U) - y,$$

the ball $B(y, s) = B_s + y$ is contained in $T(U)$. Hence $T(U)$ is an open set. $\qquad\square$

As an application of the open mapping theorem, we prove the following theorem:

**Theorem 6.4.** *Let $(X, \|\cdot\|)$ be a Banach space and $(e_n)$ a Schauder basis in $X$. For $x = \sum_{k=1}^{\infty} x_k e_k$ and $n \in \mathbf{N}$, let $P_n(x) = \sum_{k=1}^{n} x_k e_k$ be the canonical projections of the space $X$ into itself. Then operators $P_n$ as well as the coordinate functionals $x \mapsto x_k$ are bounded (cf. Exercise 5.2). Moreover, the family $\{P_n\}_{n \in \mathbf{N}}$ is norm bounded, that is, there exists a constant $C > 0$ such that $\|P_n\| < C$ for all $n \in \mathbf{N}$.*

*Proof.* Clearly, all operators $P_n$ are linear. Let $x \in X$. For $\varepsilon = 1$, there exists $N \in \mathbf{N}$ such that

$$\left\| x - \sum_{k=1}^{n} x_k e_k \right\| = \|x - P_n x\| < 1,$$

for all $n > N$. We have, for such $n$,

$$\|P_n x\| = \|P_n x - x + x\| \leq \|P_n x - x\| + \|x\| < 1 + \|x\|.$$

Hence the family of real numbers $\{\|P_n x\|\}_{n \in \mathbf{N}}$ is bounded.
We define

$$\|x\|' = \sup_{n \in \mathbf{N}} \|P_n x\| = \sup_{n \in \mathbf{N}} \left\| \sum_{k=1}^{n} x_k e_k \right\|, \qquad x \in X.$$

The function $\|\cdot\|'$ is a norm on $X$. Indeed, it is easy to verify that $\|x\|' = 0$ implies $x = 0$ and $\|\lambda x\|' = |\lambda| \|x\|'$ for every $\lambda \in \mathbf{F}$. Furthermore,

$$\|x + y\|' = \sup_{n \in \mathbf{N}} \{\|P_n x + P_n y\|\} \leq \sup_{n \in \mathbf{N}} \{\|P_n x\| + \|P_n y\|\} = \|x\|' + \|y\|',$$

that is, the triangle inequality holds.

The norm $\|\cdot\|$ is dominated by the norm $\|\cdot\|'$. Indeed, for every $\varepsilon > 0$, there exists $N \in \mathbf{N}$ such that $\|x - P_n x\| < \varepsilon$, for all $n > N$. Thus

$$\|x\| = \|P_n x + x - P_n x\| \leq \|P_n x\| + \|x - P_n x\| < \|P_n x\| + \varepsilon,$$

for all $n > N$. Suppose that $\|x\| > \sup_{n \in \mathbf{N}}\{\|P_n x\|\}$. Then there exists $\varepsilon > 0$ such that for all $n \in \mathbf{N}$, $\|x\| > \|P_n x\| + \varepsilon$, contradicting the last displayed formula. Therefore, $\|x\| \leq \|x\|'$ for all $x \in X$.

Let $(z_n)$ be a Cauchy sequence in $(X, \|\cdot\|')$. Then for a given $\varepsilon > 0$, there exists $N \in \mathbf{N}$ such that

$$\|z_m - z_n\|' < \varepsilon, \qquad \text{for all } n > m > N.$$

Because $\|z_m - z_n\| \leq \|z_m - z_n\|'$, the sequence $(z_n)$ is Cauchy in $(X, \|\cdot\|)$ and therefore converges to some $z \in X$ in the norm $\|\cdot\|$. Let $z = \sum_{k=1}^{\infty} z_k e_k$ in $(X, \|\cdot\|)$. For every $\varepsilon > 0$, there exists $N \in \mathbf{N}$ such that

$$\|z - P_n z\| < \varepsilon, \qquad \text{for all } n > N.$$

Because $P_k P_n = P_k$ for $k \leq n$ and $P_k P_n = P_n$ for $k > n$, we have

$$\|z - P_n z\|' = \sup_{k \in \mathbf{N}}\{\|P_k z - P_k P_n z\|\} = \sup_{k > n}\{\|P_k z - P_n z\|\}$$

$$= \left\| \sum_{i=n+1}^{k} z_i e_i \right\| \to 0,$$

as $n \to \infty$. Hence the series $\sum_{k=1}^{\infty} z_k e_k$ converges to $z$ in the norm $\|\cdot\|'$, that is, $(X, \|\cdot\|')$ is a Banach space.

Let $T : (X, \|\cdot\|') \to (X, \|\cdot\|)$ be the identity mapping. Clearly,

$$\|Tx\| = \|x\| \leq \|x\|'.$$

Hence $T$ is bounded. By the open mapping theorem, the inverse operator $T^{-1}$ is bounded, that is, there is a constant $C > 0$ such that

$$\|T^{-1}x\|' = \|x\|' \leq C\|x\|, \qquad \text{for all } x \in X.$$

Therefore, for all $x \in X$, $\sup_{n \in \mathbf{N}}\{\|P_n x\|\} \leq C\|x\|$, so all operators $P_n$ are bounded. Moreover, $\|P_n\| \leq C$ for all $n \in \mathbf{N}$.

We define $U_1 = P_1$ and $U_n = P_n - P_{n-1}$ for $n > 1$. It is clear that $U_n$ are bounded operators and $\|U_n\| \leq 2C$. We have

$$|x_k|\|e_k\| = \|x_k e_k\| = \|U_k x\| \leq 2C\|x\|, \qquad \text{for all } x \in X.$$

It follows that $|x_k| \leq (2C/\|e_k\|)\|x\|$, that is, the coordinate functional $x \mapsto x_k$ is bounded (cf. Exercise 5.2). $\qquad \square$

## 6.3 Closed Graph Theorem

Let $(X, \|\cdot\|_1)$ and $(Y, \|\cdot\|_2)$ be normed spaces. We define a norm $\|\cdot\|$ on the vector space $X \times Y$ by

$$\|(x, y)\| = \max\{\|x\|_1, \|y\|_2\} \tag{6.2}$$

(cf. Exercise 6.12 (a)).

If $T : X \to Y$ is a linear operator, then the graph

$$G(T) = \{(x, Tx) : x \in X\}$$

of $T$ is a subspace of the normed space $X \times Y$. (Note that from a set-theoretic point of view, $T$ and $G(T)$ are identical.) An operator $T$ is said to be *closed* if its graph is a closed subspace of $X \times Y$. If $X$ and $Y$ are Banach spaces and $T : X \to Y$ is closed, then the graph $G(T)$ of $T$ is itself a Banach space (cf. Exercise 6.12 (b), (c)).

The following is a version of the closed graph theorem.

**Theorem 6.5.** *If $X$ and $Y$ are Banach spaces and $T$ is a closed linear operator, then $T$ is bounded.*

*Proof.* We define a *projection* $\pi : G(T) \to X$ by

$$\pi(x, Tx) = x, \qquad \text{for all } x \in X.$$

It is clear that $\pi$ is a linear operator. We have

$$\|\pi(x, Tx)\|_1 = \|x\|_1 \leq \max\{\|x\|_1, \|Tx\|_2\} = \|(x, Tx)\|.$$

Hence $\pi$ is a bounded linear operator. Because $T$ is a function, $\pi$ is a bijection. By the open mapping theorem, $\pi^{-1}$ is a bounded linear operator. Therefore, there exists a constant $C > 0$ such that

$$\|\pi^{-1}x\| = \|(x, Tx)\| = \max\{\|x\|_1, \|Tx\|_2\} \leq C\|x\|_1, \qquad \text{for all } x \in X.$$

Hence for all $x \in X$, $\|Tx\|_2 \leq C\|x\|_1$, which is the desired result. $\qquad\square$

In applications of the closed graph theorem, the following equivalent version of it is useful (cf. Exercise 6.13).

**Theorem 6.6.** *Let $T : X \to Y$ be a mapping from a Banach space $X$ to a Banach space $Y$. Suppose that $T$ has the property that for every sequence $(x_n)$, if $x_n \to x$ and $Tx_n \to y$, then $y = Tx$. Then $T$ is bounded.*

The statement of the open mapping theorem follows from the closed graph theorem. We prove the following weaker version of this assertion. Namely, we use the closed graph theorem to establish the following result:

**Theorem 6.7.** *If $X$ and $Y$ are Banach spaces and $T : X \to Y$ is a bounded bijection, then $T^{-1}$ is bounded.*

We need two facts from general topology.

**Lemma 6.3.** *Let $X$ and $Y$ be topological spaces. The mapping $f : X \times Y \to Y \times X$ defined by*

$$f(x, y) = (y, x), \qquad x \in X,\ y \in Y,$$

*is a homeomorphism.*

*Proof.* Clearly, $f$ is a bijection. Let $U$ and $V$ be neighborhoods of $x \in X$ and $y \in Y$, respectively, so $V \times U$ is a neighborhood of $(y, x) \in Y \times X$. Then

$$f^{-1}(V \times U) = U \times V$$

is an open set in $X \times Y$. Because sets of the form $U \times V$ form a basis of the product topology (cf. Exercise 2.62), $f$ is continuous. By applying the same argument to $f^{-1}$, we obtain the desired result. $\qquad\square$

**Lemma 6.4.** *Suppose that $T : X \to Y$ is a continuous mapping from a topological space $X$ into a Hausdorff space $Y$. Then the graph $G(T)$ of $T$ is a closed subset of $X \times Y$.*

*Proof.* Let $(x, y)$ be a point in $X \times Y$ that does not belong to $G(T)$. Then $y \neq Tx$. Because $Y$ is Hausdorff, there are neighborhoods $U$ and $V$ of $Tx$ and $y$ such that $U \cap V = \varnothing$. Since $T$ is continuous, there is a neighborhood $W$ of $x$ such that $T(W) \subseteq U$. Hence $(z, Tz) \notin V$, for every $z \in W$. It follows that $(W \times V) \cap G(T) = \varnothing$. Because $W \times V$ is a neighborhood of $(x, y)$, the set $(X \times Y) \setminus G(T)$ is open. The result follows. $\qquad\square$

We now proceed with the proof of Theorem 6.7.

*Proof.* By Lemma 6.4, the graph $G(T)$ is a closed subspace of $X \times Y$. By Lemma 6.3, the graph of $T^{-1}$ is a homeomorphic image of $G(T)$ and therefore closed. By the closed graph theorem, $T^{-1}$ is bounded. $\qquad\square$

## 6.4 Examples and Counterexamples

The uniform boundedness theorem assumes that the space $X$ is Banach. The next example demonstrates that this assumption is essential.

*Example 6.3.* Let $X = c_{00}$ be the space of all real sequences with finite support endowed with the sup-norm, and let $(f_n)$ be a sequence of linear functionals on $X$ defined by

$$f_n(x) = nx_n, \quad \text{for } x = (x_1, \ldots, x_n, \ldots) \in c_{00}.$$

We have

$$\sup_{\|x\|=1} |f_n(x)| = \sup_{\|x\|=1} |nx_n| = n,$$

so $\|f_n\| = n$. Hence, the sequence $(\|f_n\|)$ is unbounded.

For a given $x \in X$,

$$\|f_n(x)\| = |nx_n| < \infty, \quad \text{for all } n \in \mathbf{N},$$

because $x$ has finite support. Hence the sequence $(\|f_n(x)\|)$ is bounded for every $x \in X$.

Example 6.3 shows that the result of Theorem 6.1 does not hold in general for non-Banach spaces. The space $c_{00}$ is not a Banach space, which can be shown directly by proving that the sequence

$$x^{(n)} = \left(1, \frac{1}{2}, \ldots, \frac{1}{n}, 0, 0, \ldots\right)$$

is Cauchy but diverges in $c_{00}$ (cf. Exercise 6.2). We prefer to use an indirect argument based on the observation that a proper dense subspace of a Banach space is not a Banach space (cf. Exercise 6.7). The proof of the following theorem is left as an exercise (cf. Exercise 3.21).

**Theorem 6.8.** *The space $c_{00}$ is a proper dense subspace of the space $c_0$ and therefore is not a Banach space.*

One may have a proper inclusion in Theorem 6.2 (b), as the following example illustrates.

*Example 6.4.* In the space $\ell^2$, let

$$A = \text{span}\{e_1, e_3, \ldots, e_{2k-1}, \ldots\}$$

and

$$B = \text{span}\{f_1, f_2, \ldots, f_k, \ldots\},$$

where $f_k = e_{2k} + (2k)e_{2k-1}$, $k \in \mathbf{N}$. Inasmuch as

$$e_{2k} = f_k - (2k)e_{2k-1},$$

we have $A + B = \text{span}\{e_1, \ldots, e_k, \ldots\}$; hence $\overline{A + B} = \ell^2$.

Furthermore,

$$\overline{A} = \{(x_1, x_2, \ldots) : x_{2k} = 0,\ k \in \mathbf{N},\ \sum_{j=1}^{\infty} |x_j|^2 < \infty\}$$

and

$$\overline{B} = \{(2x_2, x_2, 4x_4, x_4, \ldots, (2k)x_{2k}, x_{2k}, \ldots) : \sum_{k=1}^{\infty} ((2k)^2 + 1)|x_{2k}|^2 < \infty\}.$$

Now the vector

$$v = \left(0, \frac{1}{2}, 0, \frac{1}{4}, \ldots, 0, \frac{1}{2k}, 0, \ldots\right)$$

is in $\overline{A + B} = \ell^2$ but does not belong to $\overline{A} + \overline{B}$. Indeed, suppose that it belongs to $\overline{A} + \overline{B}$, so it is the sum of two vectors

$$(\ldots, x_{2k-1}, 0, \ldots) \in \overline{A}$$

and

$$(\ldots, (2k)y_{2k}, y_{2k}, \ldots) \in \overline{B}.$$

We have

$$x_{2k-1} + (2k)y_{2k} = 0 \quad \text{and} \quad y_{2k} = \frac{1}{2k}.$$

It follows that $x_{2k-1} = -1$ for all $k \in \mathbf{N}$, which is a contradiction. Hence $v \notin \overline{A} + \overline{B}$.

A surjective bounded mapping $T : X \to Y$ may be open if $X$ is not a Banach space, as the following example illustrates.

*Example 6.5.* In fact, a nonzero real-valued functional on every nontrivial topological vector space is an open mapping. We use the argument from the proof of Lemma 5.4. Let $f$ be a nontrivial continuous real linear functional on a topological vector space $X$ (not necessarily Banach). To show that $f$ is open, it suffices to show that $f(V)$ is open for every open neighborhood $V$ of the origin. Because $f$ is not zero, there exists $x_0 \in X$ such that $f(x_0) = c \neq 0$. We set $x^* = (1/c)x_0$, so $f(x^*) = 1$. Take $y \in f(V)$. There exists $x \in V$ such that $y = f(x)$. Because scalar multiplication is continuous, there exists $\delta > 0$ such that $x + rx^* \in V$ for all $-\delta < r < \delta$. Then $y + r \in f(V)$ for all $-\delta < r < \delta$, that is, $(y - \delta, y + \delta) \subseteq f(V)$. Hence $f(V)$ is open in $\mathbf{R}$.

The assumptions that both spaces $X$ and $Y$ in the open mapping and closed graph theorems are Banach spaces is essential. We begin with two examples illustrating this claim in the case of the space $Y$ in the open mapping theorem.

*Example 6.6.* Let $X = (\ell_1, \|\cdot\|_1)$, $Y = (\ell_1, \|\cdot\|_\infty)$, and let $T$ be the identity map from $X$ onto $Y$. Clearly, for every $x \in \ell_1$,

$$\|x\|_\infty = \sup_k |x_k| \le \sum_k |x_k| = \|x\|_1,$$

so $T$ is bounded. On the other hand, for $x^{(n)} = e_1 + \cdots + e_n$ ($\{e_k\}$ is the standard Schauder basis in $\ell_1$),

$$\|x^{(n)}\|_1 = n \qquad \text{and} \qquad \|x^{(n)}\|_\infty = 1,$$

so the inverse of $T$ is not bounded. The space $Y$ is not a Banach space. Indeed, consider the sequence

$$x^{(n)} = \left(1, \frac{1}{2}, \ldots, \frac{1}{n}, 0, 0, \ldots\right)$$

of vectors in $Y$. Suppose that $x^{(n)} \to x = (x_1, x_2, \ldots)$ in $Y$. Then

$$\|x^{(n)} - x\|_\infty = \sup_n \left\{ \sup_{1 \le k \le n} \left| x_k - \frac{1}{k} \right|, |x_{n+1}|, \ldots \right\} \to 0$$

as $n \to \infty$. It follows that $x = (1, 1/2, \ldots, 1/n, \ldots)$, which is not in $\ell_1$.

*Example 6.7.* Let $X = C[0,1]$ and

$$Y = \{x \in C^1[0,1] : x(0) = 0\},$$

both equipped with the sup-norm. The space $Y$ is not a Banach space (cf. Exercise 6.15). We define $T : X \to Y$ by

$$(Tx)(t) = \int_0^t x(u)\, du.$$

Because

$$\|Tx\| = \sup_{t \in [0,1]} \left| \int_0^t x(u)\, du \right| \le \sup_{u \in [0,1]} |x(u)| = \|x\|,$$

$T$ is bounded. The inverse operator $T^{-1} : Y \to X$ is the differentiation operator $d/dt$, which is unbounded.

It is more difficult to construct a counterexample with a non-Banach space $X$. For this we use the closed graph theorem (cf. Theorem 6.5).

*Example 6.8.* Let $X$ be an infinite-dimensional real Banach space and $\mathcal{B}$ a normalized Hamel basis in $X$. We choose a sequence $(e_n)$ in $\mathcal{B}$ and define $f(e_n) = n$ for $n \in \mathbf{N}$ and $f(e) = 0$ for $e \in \mathcal{B} \setminus \{e_n : n \in \mathbf{N}\}$. By extending $f$ linearly to the entire space $X$, we obtain an unbounded real linear functional, which we denote by the same symbol $f$.

By the closed graph theorem, the graph of $f$,

$$G = \{(x, f(x)) : x \in X\},$$

is not closed in $X \times \mathbf{R}$ in the product topology defined by the norm

$$\|(x,t)\| = \max\{\|x\|, |t|)\} \qquad \text{for } x \in X \text{ and } t \in \mathbf{R}.$$

Hence $G$ is not a Banach space. We define $T : G \to X$ by $T(x, f(x)) = x$ for $x \in X$. Clearly, $T$ is a bijection. The operator $T$ is bounded, because

$$\frac{\|x\|}{\|(x, f(x))\|} \le 1, \qquad \text{for all } x \in X, x \ne 0.$$

Inasmuch as

$$\frac{\|(x, f(x))\|}{\|x\|} \ge \frac{|f(x)|}{\|x\|}, \qquad \text{for all } x \in X, x \ne 0,$$

and $f$ is unbounded, the operator $T^{-1} : x \mapsto (x, f(x))$ is unbounded.

Now we proceed with counterexamples to the closed graph theorem. The first example (Example 6.9) shows that there exist a non-Banach space $X$ and a map $T : X \to Y$ onto a Banach space $Y$ such that the graph of $T$ is closed, whereas $T$ is unbounded.

First, we prove two lemmas.

**Lemma 6.5.** *The sequence*

$$x_n(t) = \sqrt{\left(t - \frac{1}{2}\right)^2 + \frac{1}{n}}, \qquad t \in [0, 1],$$

*converges uniformly to the function* $x(t) = |t - 0.5|$ *on* $[0, 1]$. *Hence the space* $C^1[0, 1]$ *is not complete in the sup-norm.*

*Proof.* For $\varepsilon > 0$, we have a chain of equivalent inequalities (note that $x_n(t) > x(t)$):

$$\sqrt{\left(t - \frac{1}{2}\right)^2 + \frac{1}{n}} - \left|t - \frac{1}{2}\right| < \varepsilon,$$

$$\sqrt{\left(t - \frac{1}{2}\right)^2 + \frac{1}{n}} < \left|t - \frac{1}{2}\right| + \varepsilon,$$

$$\frac{1}{n} < 2\varepsilon\left|t - \frac{1}{2}\right| + \varepsilon^2.$$

Hence, the first inequality holds for $n > 1/\varepsilon^2$ (and every $t$). $\square$

**Lemma 6.6.** *Let* $(x_n)$ *be a sequence of continuously differentiable functions on* $[0, 1]$ *such that*

*(1) $x_n \to x$ uniformly, and*
*(2) $x_n' \to y$ uniformly.*

*Then $x$ is differentiable and $x' = y$.*

*Proof.* By (2), for a given $\varepsilon > 0$, there exists $N$ such that

$$x'_n(t) - \varepsilon < y(t) < x'_n(t) + \varepsilon, \qquad t \in [0,1], \ n > N.$$

Hence for $u \in [0,1]$,

$$x_n(u) - x_n(0) - \varepsilon u \leq \int_0^u y(t)\, dt \leq x_n(u) - x_n(0) + \varepsilon u.$$

Let $h_n(u) = x_n(u) - x(u)$ and $h(u) = x(u) - x(0)$. From the last displayed inequalities we obtain

$$\left| \left[ \int_0^u y(t)\, dt - h(u) \right] - [h_n(u) - h_n(0)] \right| \leq \varepsilon u \leq \varepsilon.$$

By taking $n \to \infty$, we obtain

$$\left| \left[ \int_0^u y(t)\, dt - h(u) \right] \right| \leq \varepsilon.$$

It follows that

$$x(u) - x(0) = \int_0^u y(t)\, dt.$$

Hence $x$ is differentiable and $x' = y$. $\qquad\qquad\qquad\qquad\qquad\qquad\square$

*Example 6.9.* Let $X = C^1[0,1]$ and $Y = C[0,1]$ both with the sup-norm $\|x\|_\infty$. The space $Y$ is a Banach space, whereas $X$ is not (cf. Lemma 6.5).

The differentiation operator $T = d/dt$ maps $X$ onto $Y$. This operator is unbounded, because $\|t^n\|_\infty = 1$ and $\|nt^{n-1}\|_\infty = n$. Let $G$ be the graph of $T$ in $X \times Y$ with the product topology induced by the norm

$$\|(x,y)\| = \max\{ \sup_{t \in [0,1]} |x(t)|, \ \sup_{t \in [0,1]} |x'(t)| \}.$$

Suppose that the sequence $(x_n, x'_n)$ of vectors in $G$ converges to a vector $(x,y)$ in the space $X \times Y$. Then $x_n \to x$ and $x'_n \to y$ in the sup-norm, so $x_n$ converges uniformly to $x$ and $x'_n$ converges uniformly to $y$. Hence $y = x'$ (cf. Lemma 6.6). Therefore, $(x_n, x'_n) \to (x, x') \in G$. It follows that the graph $G$ is closed.

*Example 6.10.* Let $X$, $f$, $G$, and $T$ be as defined in Example 6.8, and $S = T^{-1}$ ($S : x \mapsto (x, f(x))$). The space $G$ is incomplete, and the operator $S : X \to G$ is unbounded (cf. Example 6.8). Let $(x_n, (x_n, f(x_n)))$ be a sequence in the graph of $S$ converging to $(x, (y, f(y)) \in X \times G$, so the sequence

$$\|(x, (y, f(y))) - (x_n, (x_n, f(x_n)))\| = \|x - x_n\| + \|(y - x_n, f(y) - f(x_n))\|$$
$$= \|x - x_n\| + \|y - x_n\| + |f(y) - f(x_n)|$$

converges to zero. It follows that $x_n \to x$ and $x_n \to y$ in $X$. Hence $y = x$, implying that the graph of $S$ is closed.

## Notes

The converse of the uniform boundedness theorem is clearly true: if a family of bounded operators from a Banach space to a normed space is norm bounded, then it is pointwise bounded.

The uniform boundedness theorem is often referred to as the "Banach–Steinhaus" theorem. The original proofs of the uniform boundedness theorem were based on a technique called the "gliding hump." Later, it was replaced by the Baire category argument. However, the "gliding hump" technique continues to be useful in functional analysis. A short proof of the uniform boundedness theorem that does not use the Baire category theorem was given by Alan D. Sokal (*American Mathematical Monthly*, May 2011, pp. 450–452).

The open mapping theorem (Theorem 6.3) was first proved by Stefan Banach in 1929 in the special case in which the operator $T$ is one-to-one. This result is known as the "bounded inverse theorem." In 1930, Juliusz Paweł Schauder found a significant simplification of Banach's proof and established the open mapping theorem in its present form. Theorem 6.4 first appeared in Banach's book (Banach 1987), originally published in Polish in 1931. This theorem plays a significant role in the theory of "classical Banach spaces" (Lindenstrauss and Tzafriri 1977).

There are many results throughout mathematics that are known by the name "closed graph theorem" (including closely related theorems). Our (functional analysis) Theorem 6.5 appeared in Banach's book (Banach 1987).

The result of Lemma 6.6 is a well-known theorem in real analysis; see, for instance, Theorem 7.12 in Wade (2000).

Examples in Section 6.4 demonstrate the crucial role of the completeness property in the proofs of the three "fundamental theorems." A similar role is played by the completeness of the field of real numbers $\mathbf{R}$ in real analysis in establishing fundamental results such as, for instance, the intermediate value theorem (cf. Ovchinnikov 2015, Section 3.7).

## Exercises

**6.1.** Explain why one can assume that the sequences $(a_n)$ in Examples 6.1 and 6.2 do not have zero terms.

**6.2.** Show that the sequence

$$x^{(n)} = \left(1, \frac{1}{2}, \dots, \frac{1}{n}, 0, 0, \dots\right)$$

of vectors in $c_{00}$ is Cauchy but does not converge to a vector in $c_{00}$.

**6.3.** Show that a proper dense subspace of a Banach space is not a Banach space.

**6.4.** Let $(T_n)$ be a sequence of bounded linear operators from a Banach space $X$ to a normed space $Y$ such that for each $x \in X$, the limit

$$\lim_{n \to \infty} T_n x$$

exists in $Y$. Prove that the operator $T : X \to Y$ defined by

$$Tx = \lim_{n \to \infty} T_n x, \qquad \text{for all } x \in X,$$

is linear and bounded.

**6.5.** Let $X$ be a Banach space, $Y$ a normed space, and $T_n : X \to Y$ a sequence of bounded operators such that $\sup\{\|T_n\| : n \in \mathbf{N}\} = \infty$. Show that there exists $x_0 \in X$ such that $\sup\{\|T_n x_0\| : n \in \mathbf{N}\} = \infty$.

**6.6.** Let $X$ be a Banach space, $Y$ a normed space, and $T_n : X \to Y$ a sequence of bounded operators such that $(T_n x)$ is Cauchy in $Y$ for every $x \in X$. Show that the sequence $(\|T_n\|)$ is bounded.

**6.7.** If $X$ and $Y$ are Banach spaces and $T_n : X \to Y$ a sequence of bounded operators, show that the following statements are equivalent:

(a) the sequence $(\|T_n\|)$ is bounded,
(b) the sequence $(\|T_n x\|)$ is bounded for every $x \in X$,
(c) the sequence $(|f(T_n x)|)$ is bounded for every $x \in X$ and every $f \in Y^*$.

**6.8.** Prove Corollary 6.1.

**6.9.** Show that:

(a) Every ball $B = B(x_0, r)$ in a normed space is a convex set.
(b) $r^{-1}(B(x_0, r) - x_0) = B(0, 1)$.

**6.10.** Let $X$ be a normed space. Prove that translation $x \mapsto x + x_0$ and dilation $x \to \lambda x$ ($\lambda \neq 0$) are homeomorphisms of $X$ onto itself.

**6.11.** Let $X$ and $Y$ be Banach spaces and $T : X \to Y$ a one-to-one bounded linear operator. Show that $T^{-1} : T(X) \to X$ is bounded if and only if $T(X)$ is closed in $Y$.

**6.12.** (a) Show that $\| \cdot \|$ defined by (6.2) is indeed a norm on $X \times Y$ and the topology defined by this norm coincides with the product topology on $X \times Y$.
(b) Prove that the product of two Banach spaces is a Banach space.
(c) Show that if $X$ and $Y$ are Banach spaces and $T : X \to Y$ is closed, then the graph $G(T)$ is a Banach space.

**6.13.** Prove Theorem 6.6.

**6.14.** Let $X$ and $Y$ be normed spaces and $T : X \to Y$ a closed linear operator.

(a) Show that the image of a compact subset of $X$ is closed in $Y$.
(b) Show that the inverse image of a compact subset of $Y$ is closed in $X$.

**6.15.** Show that the space

$$Y = \{x \in C^1[0,1] : x(0) = 0\}$$

equipped with the sup-norm is not a Banach space (cf. Lemma 6.5).

# Chapter 7
# Hilbert Spaces

Arguably, Hilbert spaces and operators on those spaces are the most popular objects of study in functional analysis because they have been very useful in many applications.

A Hilbert space is a complete inner product space, the latter being a concept well known from linear algebra. This chapter begins by introducing inner product spaces and their basic properties (cf. Section 7.1). Then we cover orthogonality of vectors and establish an important representation of a Hilbert space as a direct sum of mutually orthogonal subspaces (cf. Section 7.2). Special instances of orthonormal families of vectors in a Hilbert space are Hilbert bases, which allows for a convenient representation of vectors in the space (cf. Section 7.3). A Hilbert basis of a separable Hilbert space can be effectively constructed using the Gram–Schmidt process (cf. Theorem 7.14) for orthonormalizing a linearly independent set in an inner product space.

Fundamental tools for studying linear functionals and operators on a Hilbert space are the Riesz representation theorem for linear functionals and sesquilinear forms. This material is presented in Sections 7.4 and 7.5.

In the rest of the chapter, several classes of operators on a Hilbert space are introduced and their properties established. These are classes of normal, self-adjoint, and unitary operators (cf. Section 7.7), and the projection operators (cf. Section 7.8). The definitions of these operators use the concept of Hilbert-adjoint operator, which is the subject of Section 7.6.

## 7.1 Inner Product Spaces

**Definition 7.1.** An *inner product* on a vector space $X$ is a mapping of $X \times X$ into the scalar field $\mathbf{F}$ of $X$ that assigns to every pair $(x, y) \in X \times X$ a scalar $\langle x, y \rangle$ such that for all vectors $x, y, z \in X$ and scalars $\lambda \in \mathbf{F}$ we have the following:

© Springer International Publishing AG, part of Springer Nature 2018    149
S. Ovchinnikov, *Functional Analysis*, Universitext,
https://doi.org/10.1007/978-3-319-91512-8_7

1. $\langle x + y, z \rangle = \langle x, z \rangle + \langle y, z \rangle$.
2. $\langle \lambda x, y \rangle = \lambda \langle x, y \rangle$.
3. $\langle x, y \rangle = \overline{\langle y, x \rangle}$.
4. $\langle x, x \rangle \geq 0$ and $\langle x, x \rangle = 0$ if and only if $x = 0$.

A vector space together with an inner product on it is called an *inner product space*.

The following properties of inner products follow immediately from Definition 7.1 (cf. Exercise 7.1) and will be used below without referring to them explicitly.

1. $\langle \lambda x + \mu y, z \rangle = \lambda \langle x, z \rangle + \mu \langle y, z \rangle$,
2. $\langle x, \lambda y + \mu z \rangle = \overline{\lambda} \langle x, y \rangle + \overline{\mu} \langle x, z \rangle$,

for every $x, y, z \in X$ and $\lambda, \mu \in \mathbf{F}$. Functions satisfying these properties are known as "sesquilinear forms" (cf. Section 7.5).

*Example 7.1.* An archetypal example of an inner product space (in fact, a Hilbert space) is the space $\ell_2$ endowed with the inner product

$$\langle x, y \rangle = \sum_{k=1}^{\infty} x_k \overline{y_k}, \qquad x, y \in \ell_2. \tag{7.1}$$

By the Cauchy–Schwarz inequality (1.11), the series on the right-hand side converges absolutely, so $\langle x, y \rangle$ is well defined on $\ell_2 \times \ell_2$. The same inequality yields

$$|\langle x, y \rangle| = \left| \sum_{k=1}^{\infty} x_k \overline{y_k} \right| \leq \sum_{k=1}^{\infty} |x_k| |y_k| \leq \|x\| \|y\|. \tag{7.2}$$

It is not difficult to verify that properties 1–4 in Definition 7.1 hold for $\langle x, y \rangle$ defined by (7.1), so $\ell_2$ is an inner product space (cf. Exercise 7.2). The inner product and the norm in $\ell_2$ are related by

$$\|x\| = \sqrt{\langle x, x \rangle}, \qquad x \in \ell_2$$

(cf. (7.3) below).

*Example 7.2.* We define an inner product on the vector space $C[a, b]$ of all continuous $\mathbf{F}$-valued functions on the interval $[a, b]$ by

$$\langle x, y \rangle = \int_a^b x(t) \overline{y(t)} \, dt \qquad \text{for } x, y \in C[a, b].$$

Properties 1 and 2 of Definition 7.1 are the usual properties of the integral. For property 3, we have

$$\langle x, y \rangle = \int_a^b x(t) \overline{y(t)} \, dt = \int_a^b \overline{y(t) \overline{x(t)}} \, dt = \overline{\int_a^b y(t) \overline{x(t)} \, dt} = \overline{\langle y, x \rangle}.$$

Finally,

$$\langle x, x \rangle = \int_a^b x(t)\overline{x(t)}\, dt = \int_a^b |x(t)|^2\, dt \geq 0,$$

and $\int_a^b |x(t)|^2 dt = 0$ if and only if $x$ is the zero function on $[a, b]$ (cf. Exercise 7.3).

An inner product on a vector space $X$ defines a norm on $X$ by

$$\|x\| = \sqrt{\langle x, x \rangle}, \qquad x \in X. \tag{7.3}$$

Before proving that (7.3) indeed defines a norm on $X$ (cf. Theorem 7.2), we prove the important *Cauchy–Schwarz inequality* (cf. (7.2) in Example 7.1). This proof, as well as some other arguments below, stems from an analogue of the "square of a binomial" formula:

$$\|x + y\|^2 = \|x\|^2 + 2\mathrm{Re}\langle x, y \rangle + \|y\|^2, \tag{7.4}$$

which is also an abstract analogue of the "law of cosines" in geometry.
    Indeed, we have

$$\begin{aligned}
\|x + y\|^2 &= \langle x + y, x + y \rangle = \langle x, x \rangle + \langle x, y \rangle + \langle y, x \rangle + \langle y, y \rangle \\
&= \langle x, x \rangle + \langle x, y \rangle + \overline{\langle x, y \rangle} + \langle y, y \rangle \\
&= \|x\|^2 + 2\mathrm{Re}\langle x, y \rangle + \|y\|^2.
\end{aligned}$$

**Theorem 7.1. (Cauchy–Schwarz inequality.)** *If $X$ is an inner product space and $\|x\|$ is defined by (7.3), then*

$$|\langle x, y \rangle| \leq \|x\|\|y\|. \tag{7.5}$$

*Moreover, equality holds in (7.5) if and only if one of $x, y$ is a scalar multiple of the other.*

*Proof.* We may assume that $x \neq 0$ and $y \neq 0$, because otherwise (7.5) trivially holds (cf. Exercise 7.5) and the vectors are linearly dependent. Note that then (7.5) is equivalent to

$$\left| \left\langle \frac{x}{\|x\|}, \frac{y}{\|y\|} \right\rangle \right| \leq 1.$$

We define $u = x/\|x\|$ and $v = y/\|y\|$. Then by (7.4),

$$\begin{aligned}
0 &\leq \|u - \langle u, v \rangle v\|^2 \\
&= \|u\|^2 - 2\mathrm{Re}\langle u, \langle u, v \rangle v \rangle + |\langle v, u \rangle|^2 \|v\|^2 \\
&= 1 - 2\mathrm{Re}(\overline{\langle u, v \rangle}\langle u, v \rangle) + |\langle u, v \rangle|^2 \\
&= 1 - 2|\langle u, v \rangle|^2 + |\langle u, v \rangle|^2 = 1 - |\langle u, v \rangle|^2.
\end{aligned} \tag{7.6}$$

Hence
$$|\langle u, v \rangle| \le 1,$$
or equivalently,
$$\frac{|\langle x, y \rangle|}{\|x\|\|y\|} \le 1,$$

which is the desired inequality. Equality (7.5) holds if and only if it holds in (7.6), that is, if $u - \langle u, v \rangle v = 0$ (cf. (7.3)), so the vector $u$ is a scalar multiple of the vector $v$. From this, the second statement of the theorem easily follows.                                                                    □

**Theorem 7.2.** *Formula (7.3) defines a norm on the inner product space $X$.*

*Proof.* We need to verify that

(a)    $\|\lambda x\| = |\lambda|\|x\|,$                           *homogeneity*
(b)    $\|x + y\| \le \|x\| + \|y\|,$                          *triangle inequality*
(c)    $\|x\| = 0$   implies   $x = 0$

(cf. Definition 3.2). The homogeneity of $\|\cdot\|$ is the result of Exercise 7.6, and part (c) follows immediately from the definition of the inner product.

It remains to establish the triangle inequality. By (7.4) and the Cauchy–Schwarz inequality, we obtain

$$\begin{aligned}
\|x + y\|^2 &= \|x\|^2 + 2\mathrm{Re}\langle x, y \rangle + \|y\|^2 \\
&\le \|x\|^2 + 2|\langle x, y \rangle| + \|y\|^2 \\
&\le \|x\|^2 + 2\|x\|\|y\| + \|y\|^2 = (\|x\| + \|y\|)^2,
\end{aligned}$$

because the real part of a complex number cannot exceed its absolute value.
                                                                    □

The Cauchy–Schwarz inequality is a very useful inequality, and it is encountered in many different settings. It will be used several times later in this chapter. An elegant use of this inequality is found in the following example.

*Example 7.3.* For $0 < t < 1$, let $C[0, t]$ be the real inner product space from Example 7.2 and let $T : C[0, 1] \to C[0, 1]$ be the indefinite integration operator (cf. Example 4.6)

$$(Tx)(t) = \int_0^t x(s)\,ds, \qquad t \in [0, 1].$$

Below, we use the Cauchy–Schwarz inequality (7.5) for $C[0, t]$ in the second line and Fubini's theorem in the fourth and fifth lines:

$$\|Tx\|^2 = \int_0^1 \left| \int_0^t x(s)\, ds \right|^2 dt = \int_0^1 \left| \int_0^t \sqrt{\cos \frac{\pi}{2}s} \cdot \frac{x(s)}{\sqrt{\cos \frac{\pi}{2}s}}\, ds \right|^2 dt$$

$$\leq \int_0^1 \left( \int_0^t \cos \frac{\pi}{2}s\, ds \cdot \int_0^t \frac{|x(s)|^2}{\cos \frac{\pi}{2}s}\, ds \right) dt$$

$$= \frac{2}{\pi} \int_0^1 \left( \sin \frac{\pi}{2}t \cdot \int_0^t \frac{|x(s)|^2}{\cos \frac{\pi}{2}s}\, ds \right) dt$$

$$= \frac{2}{\pi} \int_0^1 \int_0^t \left( \sin \frac{\pi}{2}t \cdot \frac{|x(s)|^2}{\cos \frac{\pi}{2}s} \right) ds\, dt$$

$$= \frac{2}{\pi} \int_0^1 \int_s^1 \left( \sin \frac{\pi}{2}t \cdot \frac{|x(s)|^2}{\cos \frac{\pi}{2}s} \right) dt\, ds$$

$$= \frac{2}{\pi} \cdot \frac{2}{\pi} \int_0^1 \cos \frac{\pi}{2}s \cdot \frac{|x(s)|^2}{\cos \frac{\pi}{2}s}\, ds = \left( \frac{2}{\pi} \right)^2 \|x\|^2.$$

By Theorem 7.1, we have equality in the second line if $x(s) = \lambda \cos \frac{\pi}{2}s$ $(\lambda \neq 0)$. Hence $\|T\| = 2/\pi$.

We have proved (cf. Theorem 7.2) that every inner product space is also a normed space if the norm is defined by (7.3). Here is a natural question: Given a norm on a vector space, can we find an inner product on that space that will determine this norm by means of (7.3)? To address this question, we use the *parallelogram equality*:

$$\|x + y\|^2 + \|x - u\|^2 = 2(\|x\|^2 + \|y\|^2) \tag{7.7}$$

for all vectors $x$ and $y$ in an inner product space $X$. The proof is very straight-forward. Use the identity (7.4) with $y$ replaced by $-y$ to obtain the identity

$$\|x - y\|^2 = \|x\|^2 - 2\mathrm{Re}\langle x, y \rangle + \|y\|^2. \tag{7.8}$$

Then add these identities to obtain (7.7).

In the following examples, we use the parallelogram equality to show that there are normed spaces with norms not induced by inner products.

*Example 7.4.* Consider the Banach spaces $c$, $c_0$, $\ell_\infty$, and $\ell_p$ for $p \neq 2$. It is clear that the vectors

$$x = (1, 1, 0, 0, \ldots) \quad \text{and} \quad y = (1, -1, 0, 0, \ldots)$$

belong to these spaces. We have

$$x + y = (2, 0, 0, 0, \ldots) \quad \text{and} \quad x - y = (0, 2, 0, 0, \ldots).$$

The supremum and $p$-norms of these vectors are

$$\|x\|_\infty = \|y\|_\infty = 1 \quad \text{and} \quad \|x + u\|_\infty = \|x - y\|_\infty = 2$$

and
$$\|x\|_p = \|y\|_p = 2^{1/p} \quad \text{and} \quad \|x+y\|_p = \|x-y\|_p = 2,$$
respectively. We have
$$\|x+y\|_\infty^2 + \|x-y\|_\infty^2 = 8 \quad \text{and} \quad 2(\|x\|_\infty^2 + \|y\|_\infty^2) = 4 \neq 8,$$
and
$$\|x+y\|_p^2 + \|x-y\|_p^2 = 8 \quad \text{and} \quad 2(\|x\|_p^2 + \|y\|_p^2) = 4 \cdot 2^{1/p} \neq 8.$$

Because the parallelogram identity does not hold in the spaces $c$, $c_0$, $\ell_\infty$, and $\ell_p$ ($p \neq 2$), these spaces are not inner product spaces.

Note that for $p = 2$, the space $\ell_2$ is an inner product space (cf. Example 7.1).

*Example 7.5.* We show that the vector space $C[0,1]$ endowed with the sup-norm $\|x\|_\infty$ is not an inner product space. For functions $x(t) = 1$ and $y(t) = t$ on $[0,1]$, we have
$$\|x\|_\infty = 1 \quad \text{and} \quad \|y\|_\infty = 1,$$
and
$$\|x+y\|_\infty = \sup_{t \in [0,1]} (1+t) = 2 \quad \text{and} \quad \|x-y\|_\infty = \sup_{t \in [0,1]} (1-t) = 1.$$

Hence
$$\|x+y\|_\infty^2 + \|x-y\|_\infty^2 = 5 \neq 4 = 2(\|x\|_\infty^2 + \|y\|_\infty^2),$$
so the parallelogram identity does not hold.

Suppose we know the norm on an inner product space. The next theorem shows how the inner product can be "recaptured" from the norm.

**Theorem 7.3. (Polarization identities.)** *If $X$ is a real inner product space, then*
$$\langle x, y \rangle = \tfrac{1}{4}\left(\|x+y\|^2 - \|x-y\|^2\right), \qquad x, y \in X.$$

*For a complex inner product space $X$, we have*
$$\langle x, y \rangle = \tfrac{1}{4}\left(\|x+y\|^2 - \|x-y\|^2 + i\|x+iy\|^2 - i\|x-iy\|^2\right), \qquad x, y \in X.$$

*Proof.* By subtracting the identity (7.8) from the identity (7.4), we obtain
$$4\mathrm{Re}\langle x, y \rangle = \|x+y\|^2 - \|x-y\|^2. \tag{7.9}$$

This is the polarization identity in the case of a real space $X$, because $\mathrm{Re}\langle x, y \rangle = \langle x, y \rangle$.

Inasmuch as $\langle x, y \rangle = \mathrm{Re}\langle x, y \rangle + i\,\mathrm{Im}\langle x, y \rangle$ in the complex case, we have
$$\langle x, y \rangle = \mathrm{Re}\langle x, y \rangle + i\mathrm{Re}\langle x, iy \rangle,$$

because

$$Im\langle x, y \rangle = Re(-i\langle x, y \rangle) = Re\langle x, iy \rangle.$$

By applying (7.9), we obtain the polarization identity in the complex case:

$$4\langle x, y \rangle = 4Re\langle x, y \rangle + i4Re\langle x, iy \rangle$$
$$= \|x + y\|^2 - \|x - y\|^2 + i\|x + iy\|^2 - i\|x - iy\|^2.$$

The result follows.                                                                    □

An important consequence of Theorem 7.3 is the fact that norms determine the inner product in inner product spaces.

**Definition 7.2.** Inner product spaces $X$ and $Y$ are said to be *isomorphic* if there is a bijective linear map $T : X \to Y$ that preserves the inner products, that is, $\langle Tx, Tx' \rangle = \langle x, x' \rangle$ for all $x, x' \in X$.

Theorem 7.3 tells us that it suffices to require that the bijective linear map $T : X \to Y$ be an isometry of $X$ onto $Y$ to show that these inner product spaces are isomorphic.

The continuity properties of the inner product are the subject of the next theorem.

**Theorem 7.4.** *If $(x_n)$ and $(y_n)$ are Cauchy sequences in an inner product space $X$, then the sequence $(\langle x_n, y_n \rangle)$ converges.*

*Furthermore, if $x_n \to x$ and $y_n \to y$ in $X$, then $\langle x_n, y_n \rangle \to \langle x, y \rangle$.*

*Proof.* Using the Cauchy–Schwarz inequality, we obtain

$$|\langle x_n, y_n \rangle - \langle x_m, y_m \rangle| = |\langle x_n, y_n \rangle - \langle x_n, y_m \rangle + \langle x_n, y_m \rangle - \langle x_m, y_m \rangle|$$
$$\leq |\langle x_n, y_n \rangle - \langle x_n, y_m \rangle| + |\langle x_n, y_m \rangle - \langle x_m, y_m \rangle|$$
$$= |\langle x_n, y_n - y_m \rangle| + |\langle x_n - x_m, y_m \rangle|$$
$$\leq \|x_n\|\|y_n - y_m\| + \|x_n - x_m\|\|y_m\|.$$

Because the sequences $(x_n)$ and $(y_n)$ are Cauchy and bounded (cf. Theorem 2.7), the sequence of scalars $(\langle x_n, y_n \rangle)$ is Cauchy and therefore convergent.

If $\|x_n - x\| \to 0$ and $\|y_n - y\| \to 0$, then as before,

$$0 \leq |\langle x_n, y_n \rangle - \langle x, y \rangle| = |\langle x_n, y_n \rangle - \langle x_n, y \rangle + \langle x_n, y \rangle - \langle x, y \rangle|$$
$$\leq |\langle x_n, y_n \rangle - \langle x_n, y \rangle| + |\langle x_n, y \rangle - \langle x, y \rangle|$$
$$= |\langle x_n, y_n - y \rangle| + |\langle x_n - x, y \rangle|$$
$$\leq \|x_n\|\|y_n - y\| + \|x_n - x\|\|y\| \to 0,$$

because the convergent sequence $(x_n)$ is bounded.                                     □

**Definition 7.3.** A *Hilbert space* is a complete inner product space.

A subspace $Y$ of an inner product space $X$ is a vector subspace with the inner product restricted to $Y$. A closed subspace of a Hilbert space is a Hilbert space (cf. Exercise 7.8).

In the rest of this section, we prove that every inner product space can be completed. The resulting completion is a Hilbert space, which is unique up to isomorphism.

**Theorem 7.5.** *Let $X$ be an inner product space. Then there exist a Hilbert space $H$ and an isometry $T : X \to H$ such that $T(X)$ is dense in $H$. The space $H$ is unique up to isomorphisms of inner product spaces.*

*Proof.* By Theorem 4.2, we can always complete the normed space $X$ to obtain its Banach completion, which we denote by $H$. The image $W = T(X)$ is a normed space that is dense in $H$ and isometric to $X$ as a normed space. We define an inner product on $W$ by $\langle Tx, Ty \rangle = \langle x, y \rangle$ for $x, y \in X$. By the polarization identities (cf. Theorem 7.3), $W$ is isomorphic to $X$ as an inner product space.

For $x$ and $y$ in $H$, let $(x_n)$ and $(y_n)$ be sequences in $W$ converging to $x$ and $y$, respectively. By Theorem 7.4, the sequence $(\langle x_n, y_n \rangle)$ is convergent. We define

$$\langle x, y \rangle = \lim_{n \to \infty} \langle x_n, y_n \rangle \tag{7.10}$$

and show that this limit does not depend on the choice of sequences $(x_n)$ and $(y_n)$ converging to $x$ and $y$, respectively. Suppose that $(x_n')$ and $(y_n')$ are other sequences converging to $x$ and $y$, respectively. By the previous argument, the sequence $(\langle x_n', y_n' \rangle)$ is convergent. We have

$$0 \le |\langle x_n, y_n \rangle - \langle x_n', y_n' \rangle| \le |\langle x_n, y_n \rangle - \langle x_n', y_n \rangle| + |\langle x_n', y_n \rangle - \langle x_n', y_n' \rangle|$$
$$= |\langle x_n - x_n', y_n \rangle| + |\langle x_n', y_n - y_n' \rangle|$$
$$\le \|x_n - x_n'\| \|y_n\| + \|x_n'\| \|y_n - y_n'\| \to 0.$$

It follows that $\lim_{n \to \infty} \langle x_n', y_n' \rangle = \lim_{n \to \infty} \langle x_n, y_n \rangle$, so $\langle x, y \rangle$ in (7.10) is well defined.

We leave it to the reader to show that (7.10) is an inner product on the space $H$ (cf. Exercise 7.7). Now, for $(x_n) \to x$, where $(x_n) \in W$ and $x \in H$, we have

$$\|x\| = \lim_{n \to \infty} \|x_n\| = \lim_{n \to \infty} \sqrt{\langle x_n, x_n \rangle} = \sqrt{\langle x, x \rangle}.$$

Therefore, the Banach space $H$ is a Hilbert space.

Finally, suppose that a Hilbert space $H'$ is another completion of the space $X$. By Theorem 4.2, the Banach spaces $H$ and $H'$ are isometric. By the polarization identities, the Hilbert spaces $H$ and $H'$ are isomorphic.  $\square$

## 7.2 Orthogonality

Let us recall the definition of orthogonality that was introduced at the end of Section 4.4. In a normed space $X$, two vectors $x$ and $y$ are said to be orthogonal if $\|x - ky\| \geq \|x\|$ for all scalars $k \in \mathbf{F}$. We write $x \perp y$, for two orthogonal vectors $x$ and $y$.

In the case of an inner product space, we us a different definition that also stems from geometry.

**Definition 7.4.** In an inner product space, vectors $x$ and $y$ are said to be *orthogonal* if $\langle x, y \rangle = 0$. The same notation $x \perp y$ is used for two orthogonal vectors $x$ and $y$.

In fact, it is not difficult to prove that in an inner product space, the two definitions of orthogonality are equivalent.

**Theorem 7.6.** *In an inner product space, $\langle x, y \rangle = 0$ if and only if*

$$\|x - ky\| \geq \|x\|$$

*for all scalars $k \in \mathbf{F}$.*

*Proof.* (Necessity.) Suppose that $\langle x, y \rangle = 0$. Then for every $k$,

$$\|x - ky\|^2 = \|x\|^2 - 2\mathrm{Re}(\overline{k}\langle x, y \rangle) + \|ky\|^2 = \|x\|^2 + \|ky\|^2 \geq \|x\|^2.$$

Hence $\|x - ky\| \geq \|x\|$.

(Sufficiency.) Suppose that $\|x - ky\| \geq \|x\|$ for all scalars $k$. Then for $t \in \mathbf{R}$,

$$\|x\|^2 \leq \|x - ty\|^2 = \|x\|^2 - 2t\mathrm{Re}\langle x, y \rangle + t^2\|y\|^2.$$

Therefore,

$$2t\mathrm{Re}\langle x, y \rangle \leq t^2\|y\|^2.$$

If $\mathrm{Re}\langle x, y \rangle \geq 0$, choose $t > 0$, to obtain

$$\mathrm{Re}\langle x, y \rangle \leq \frac{t}{2}\|y\|^2,$$

and for $\mathrm{Re}\langle x, y \rangle < 0$, choose $t < 0$, and obtain

$$-\mathrm{Re}\langle x, y \rangle \leq \frac{|t|}{2}\|y\|^2.$$

Because $t$ is an arbitrary nonzero number, $\mathrm{Re}\langle x, y \rangle = 0$. Similarly, it follows that $\mathrm{Im}\langle x, y \rangle = \mathrm{Re}\langle x, iy \rangle$ also equals zero, so $\langle x, y \rangle = 0$. $\square$

In the rest of this chapter, we use Definition 7.4.

*Example 7.6.* A simple argument shows that unless an inner product space $X$ is one-dimensional (cf. Exercise 7.16), there are many pairs of mutually orthogonal nonzero vectors in $X$. Indeed, let $x, y \in X$ be linearly independent vectors. We want to find a scalar $\lambda$ such that the vector $z = x - \lambda y$ is orthogonal to $y$. For this, we need

$$\langle z, y \rangle = \langle x, y \rangle - \lambda \langle y, y \rangle = 0.$$

By setting $\lambda = \langle x, y \rangle / \|y\|^2$ (note that $y \neq 0$), we obtain $z \perp y$. In particular, if $\|y\| = 1$, then

$$z = x - \langle x, y \rangle y \quad \text{is orthogonal to} \quad y.$$

It is clear that $\text{span}\{z, y\} = \text{span}\{x, y\}$.

The result of this example will be used later in the chapter.

We use the following notation. Let $X$ be an inner product space. If $x \in X$ is orthogonal to every element of a nonempty subset $E$ of $X$, then we say that $x$ is orthogonal to $E$ and write $x \perp E$. If $E$ and $F$ are nonempty subsets of $X$, the notation $E \perp F$ means that $x \perp y$ for all $x \in E$ and $y \in F$. Also, $E^\perp = \{y \in X : y \perp x, \text{ for every } x \in E\}$. The set $E^\perp$ is called the *orthogonal complement* of $E$.

We now proceed with proving the "direct sum decomposition theorem" (Theorem 7.8 below), which is an important tool in the study of linear operators. First, we establish the existence of the projection of a vector in a Hilbert space onto a closed subspace.

**Theorem 7.7.** *Let $E$ be a closed proper subspace of a Hilbert space $H$, and $x$ a vector in $H \setminus E$. Then there exists a unique vector $y \in E$ that is closest to $x$ in the sense that*

$$\|x - y\| = d(x, E) = \inf_{u \in E} \|x - u\|.$$

*This vector $y$ is called the* projection *of $x$ onto $E$.*

*Proof.* Let $d = d(x, E)$. By the definition of infimum, there is a sequence $(y_n)$ of vectors in $E$ such that $\|x - y_n\| \to d$. We apply the parallelogram identity (cf. (7.7)) to the vectors $x - y_n$ and $x - y_m$,

$$\|2x - y_n - y_m\|^2 + \|y_n - y_m\|^2 = 2\|x - y_n\|^2 + 2\|x - y_m\|^2,$$

to obtain

$$\|y_n - y_m\|^2 = 2\|x - y_n\|^2 + 2\|x - y_m\|^2 - \|2x - y_n - y_m\|^2. \tag{7.11}$$

For $\varepsilon > 0$, there exists $N \in \mathbf{N}$ such that

$$\|x - y_m\|^2 < d^2 + \varepsilon \qquad \text{and} \qquad \|x - y_n\|^2 < d^2 + \varepsilon$$

for all $m, n > N$. Furthermore,

$$\|2x - y_n - y_m\|^2 = 4\left\|x - \frac{y_n + y_m}{2}\right\|^2 \geq 4d^2,$$

because $(y_n + y_m)/2 \in E$. Therefore, by (7.11),

$$\|y_n - y_m\|^2 < 2(d^2 + \varepsilon) + 2(d^2 + \varepsilon) - 4d^2 = 4\varepsilon, \qquad \text{for } m, n > N.$$

It follows that $(y_n)$ is a Cauchy sequence in $E$ that converges to some $y \in E$, since $E$ is a closed subspace and therefore complete. By the continuity of the norm,

$$\|x - y\| = \lim_{n \to \infty} \|x - y_n\| = d.$$

This proves the existence of a closest point.

The uniqueness of $y$ follows again from the parallelogram identity. Suppose that $\|x - y\| = \|x - y'\| = d$. Then, as in (7.11),

$$\|y - y'\|^2 = 2\|x - y\|^2 + 2\|x - y'\|^2 - \|2x - y - y'\|^2 \leq 4d^2 - 4d^2 = 0.$$

Hence $y' = y$. $\qquad\qquad\qquad\qquad\qquad\qquad\qquad\qquad\qquad\qquad\qquad\qquad\qquad$ □

*Example 7.7.* Consider the real Banach space $H = \ell_\infty^2$ and the closed subspace $E = \{(x, 0) : x \in \mathbf{R}\}$ of $H$. The distance from the vector $(0, 1)$ to a vector $(x, 0) \in E$ is $\max\{|x|, 1\} \geq 1$. Hence all vectors $\{(x, 0) \in E : |x| \leq 1\}$ are closest to $(0, 1)$.

Note that $\ell_\infty^2$ is not an inner product space (cf. Example 7.4).

The statements of Theorems 7.7 and 7.8 are illustrated by the drawing in Fig. 7.1.

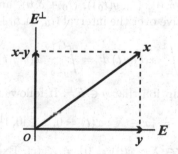

**Fig. 7.1** Orthogonal decomposition.

**Theorem 7.8.** *If $E$ is a closed subspace of a Hilbert space $H$, then*

$$H = E \oplus E^\perp.$$

*Proof.* By the result of Exercise 7.19(c), $E^\perp$ is a closed subspace of $H$. If $x \in E$ and $x \in E^\perp$, then $\langle x, x \rangle = 0$. Hence $x = 0$. Therefore, $E \cap E^\perp = \{0\}$.

For $x \in H$, let $x_1 \in E$ be its projection onto $E$ (cf. Theorem 7.7). Then

$$\|x - x_1\| \le \|x - (x_1 + kv)\|, \qquad \text{for every } v \in E \text{ and } k \in \mathbf{F},$$

because $x_1 - kv \in E$. Put $x_2 = x - x_1$. Then

$$\|x_2\| \le \|x_2 - kv\|, \qquad \text{for every } v \in E \text{ and } k \in \mathbf{F}.$$

By Theorem 7.6, $x_2 \perp v$. Because $v$ is an arbitrary vector in $E$, the vector $x_2$ belongs to $E^\perp$. Since $x = x_1 + x_2$, we have proved that $H = E \oplus E^\perp$. $\qquad \square$

The following example shows that Theorem 7.8 does not hold in general for incomplete inner product spaces (cf. Exercise 7.20).

*Example 7.8.* Let $X$ be the space $C[-1, 1]$ of real continuous functions on $[-1, 1]$ with the inner product defined by

$$\langle x, y \rangle = \int_{-1}^{1} x(t)y(t)\, dt, \qquad \text{for } x, y \in C[-1, 1]$$

(cf. Example 7.2). Consider the subspace

$$E = \{x \in X : x(t) = 0,\ t \in [-1, 0]\}$$

of $X$. This subspace is closed (cf. Exercise 7.20).

Let $y$ be a nonzero function in $E^\perp$. We may assume that there exists $t_0 \in (0, 1)$ such that $y(t_0) > 0$. Because $y$ is a continuous function, there exists $\delta > 0$ such that $y(t) > 0$ on $[t_0 - \delta, t_0 + \delta] \subseteq (0, 1)$.

Define $x \in E$ as a continuous piecewise linear function on $[-1, 1]$ with nodes at $(-1, 0)$, $(t_0 - \delta, 0)$, $(t_0, y(t_0))$, $(t_0 + \delta, 0)$, and $(1, 0)$ (cf. Fig. 7.2). Clearly, $x(t)y(t)$ is positive over the interval $(t_0 - \delta, t_0 + \delta)$ and zero otherwise. We have

$$\langle x, y \rangle = \int_{-1}^{1} x(t)y(t)\, dt = \int_{t_0 - \delta}^{t_0 + \delta} x(t)y(t)\, dt > 0,$$

contradicting our assumption that $y \in E^\perp$. It follows that

$$E^\perp = \{x \in X : x(t) = 0,\ t \in [0, 1]\}.$$

Because $E \oplus E^\perp = \{x \in X : x(0) = 0\} \ne X$ (cf. Exercise 7.20), the direct decomposition theorem does not hold for the subspace $E$.

Let $E$ be a nonempty subset of a Hilbert space $H$. Because $\perp$ (orthogonality) is a symmetric binary relation on $H$, every element of $E$ is orthogonal to $E^\perp$. Hence, $E \subseteq (E^\perp)^\perp = E^{\perp\perp}$. If $E$ is a closed subspace of $H$, then by Theorem 7.8,

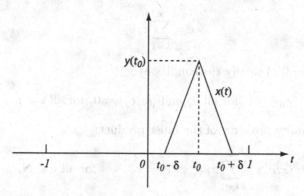

**Fig. 7.2** The function $x$ in Example 7.8.

$$H = E \oplus E^{\perp} = E^{\perp} \oplus E^{\perp\perp}.$$

Hence $E$ cannot be a proper subspace of $E^{\perp\perp}$. We established the following corollary of Theorem 7.8.

**Corollary 7.1.** *If $E$ is a closed subspace of a Hilbert space $H$, then $E^{\perp\perp} = E$.*

For a nonempty subset $S$ of a Hilbert space $H$, its *closed span* is the set $\overline{\text{span}}\,S$, which is a closed subspace of $H$. Let $E = \overline{\text{span}}\,S$. Because $S \subseteq S^{\perp\perp}$, and $S^{\perp\perp}$ is a closed subspace of $H$, it follows that $E \subseteq S^{\perp\perp}$. On the other hand, $S \subseteq E$ clearly implies $E^{\perp} \subseteq S^{\perp}$, which in turn implies, by Corollary 7.1,

$$S^{\perp\perp} \subseteq E^{\perp\perp} = E.$$

We have proved another corollary:

**Corollary 7.2.** *For a nonempty subset $S$ of a Hilbert space $H$, its closed span is $S^{\perp\perp}$.*

## 7.3 Orthonormal Sets

*Example 7.9.* We showed in Example 7.1 that the Banach space $\ell_2$ with the inner product defined by

$$\langle x, y \rangle = \sum_{k=1}^{\infty} x_k \overline{y}_k,$$

where $x = (x_n)$ and $y = (y_n)$ are vectors in $\ell_2$, is an inner product space. Let $(e_n)$ be the standard Schauder basis for $\ell_2$ (cf. Exercise 4.3). Every vector $x = (x_n)$ in the space $\ell_2$ admits the following expansion:

$$x = \sum_{k=1}^{\infty} x_k e_k.$$

The vectors in $(e_n)$ satisfy the conditions

$$\|e_n\| = 1, \text{ for all } n, \text{ and } \langle e_i, e_j \rangle = 0, \text{ for all } i \neq j.$$

By the continuity property of the inner product,

$$\langle x, e_n \rangle = \sum_{k=1}^{\infty} x_k \langle e_k, e_n \rangle = x_n, \qquad \text{for all } n \in \mathbf{N}.$$

Therefore,

$$\sum_{k=1}^{\infty} |\langle x, e_k \rangle|^2 = \|x\|^2 \quad \text{and} \quad x = \sum_{k=1}^{\infty} \langle x, e_k \rangle e_k.$$

These properties of the space $\ell_2$ motivate most discussions in this section.

**Definition 7.5.** Let $J$ be a nonempty set of indices. A family $\{e_i\}_{i \in J}$ of vectors in an inner product space $X$ is said to be *orthogonal* if $\langle e_i, e_j \rangle = 0$ for all $i \neq j$ in $J$, and it is said to be *orthonormal* if in addition, $\|e_i\| = 1$ for all $i \in J$.

If the family $\{e_i\}_{i \in J}$ is orthonormal, then the mapping $i \mapsto e_i$ is one-to-one, so we can talk about the *orthonormal set* $\{e_i : i \in J\}$.

*Example 7.10.* The vectors of the Schauder basis $(e_n)$ for the space $\ell_2$ form an orthonormal family (cf. Example 7.9).

*Example 7.11.* Let $X$ be the vector space $C[-\pi, \pi]$ of real continuous functions on the interval $[-\pi, \pi]$ with the inner product defined by

$$\langle x, y \rangle = \int_{-\pi}^{\pi} x(t)y(t)\, dt, \qquad \text{for } x, y \in C[-\pi, \pi]$$

(cf. Example 7.8). The family

$$\left\{ \frac{1}{\sqrt{\pi}} \sin(nt), \quad \frac{1}{\sqrt{\pi}} \cos(nt), \quad \frac{1}{\sqrt{2\pi}} \right\}_{n \in \mathbf{N}}$$

is orthonormal (cf. Exercise 7.24).

An immediate property of an orthonormal family of vectors is the result of the following lemma.

**Lemma 7.1.** *An orthonormal family* $\{e_i\}_{j \in J}$ *of vectors in an inner product space* $X$ *is linearly independent.*

*Proof.* Consider the equation

$$\sum_{i \in A} \lambda_i e_i = 0,$$

where $A$ is a finite subset of $J$ and $\{\lambda_i\}_{i \in A}$ is a family of scalars. For every $j \in A$ we have

$$0 = \left\langle \sum_{i \in A} \lambda_i e_i, e_j \right\rangle = \sum_{i \in A} \lambda_i \langle e_i, e_j \rangle = \lambda_j.$$

The result follows, because $A$ is an arbitrary finite subset of $J$.                    □

Note that an orthogonal family is linearly dependent if it contains the zero vector.

We adopt the following definition of *summability* of a family of vectors in a normed space. (See also Section A.1.)

**Definition 7.6.** A family of vectors $\{x_i\}_{i \in J}$ in a normed space $X$ is said to be *summable with sum* $x \in X$ if for every $\varepsilon > 0$, there exists a finite subset $A$ of the set $J$ such that

$$\left\| x - \sum_{i \in B} x_i \right\| < \varepsilon$$

for every finite set $B$ with $A \subseteq B \subseteq J$. If $\{x_i\}_{i \in J}$ is summable with sum $x$, we write $\sum_{i \in J} x_i = x$.

We can apply this definition to the one-dimensional real normed space $\mathbf{R}$ to obtain the concept of a summable family of real numbers. The following lemma gives a convenient criterion for summability of families of nonnegative numbers.

**Lemma 7.2.** *Let $\{a_i\}_{i \in J}$ be a family of nonnegative numbers such that all finite partial sums are bounded by the same number $M$, that is,*

$$\sum_{i \in A} a_i \le M, \qquad \text{for all finite sets } A \subseteq J.$$

*Then the family $\{a_i\}_{i \in J}$ is summable with sum $a = \sup_A \sum_{i \in A} a_i$, where the supremum is taken over the family of all finite subsets of $J$. Furthermore,*

$$\sum_{i \in J} a_i \le M.$$

*Proof.* By the definition of supremum, for every $\varepsilon > 0$, there is a finite subset $A$ of $J$ such that

$$a - \sum_{i \in A} a_i < \varepsilon.$$

Let $B$ be a finite subset of $J$ such that $A \subseteq B$. Inasmuch as the numbers $a_i$ are nonnegative, we have $\sum_{i \in A} a_i \le \sum_{i \in B} a_i$. Hence

$$a - \sum_{i \in B} a_i \le a - \sum_{i \in A} a_i < \varepsilon.$$

The result follows. $\qquad\square$

The following property of families of nonnegative numbers is known as the "principle of comparison." The proof of this property is left to the reader (cf. Exercise 7.25).

**Lemma 7.3.** *Let $\{a_i\}_{i \in J}$ and $\{b_i\}_{i \in J}$ be two families of nonnegative numbers such that $a_i \le b_i$ for all $i \in J$. If the family $\{b_i\}_{i \in J}$ is summable, then so is $\{a_i\}_{i \in J}$, and*

$$\sum_{i \in J} a_i \le \sum_{i \in J} b_i.$$

It can be shown that for a countable set $J$ and every enumeration of its elements $J = \{i_1, i_2, \ldots\}$, one has

$$\sum_{j \in J} a_i = \sum_{k=1}^{\infty} a_{i_k}$$

for a summable family of nonnegative numbers $\{a_i\}_{i \in J}$ (cf. Exercise 7.28). (This follows from the "rearrangement theorem" for absolutely convergent series in real analysis.)

**Lemma 7.4.** *For an arbitrary set $J$, there are at most countably many nonzero elements in the summable family of nonnegative numbers $\{a_i\}_{i \in J}$.*

*Proof.* Indeed, let $J' = \{i \in J : a_i > 0\}$. Then $J' = \bigcup_{k=1}^{\infty} J_k$, where

$$J_k = \{i \in J : a_i > 1/k\}.$$

Clearly, each set $J_k$ is finite. Hence $J'$ is at most countable. $\qquad\square$

We will need another property of summable families of scalars. Informally, it shows that zero terms do not contribute to the sum of a summable family.

**Lemma 7.5.** *Let $\{a_i\}_{i \in J}$ be a family of numbers and $J_0 = \{i \in J : a_i \ne 0\}$. If the family $\{a_i\}_{i \in J_0}$ is summable with sum $a$, then $\{a_i\}_{i \in J}$ is summable with the same sum.*

*Proof.* For $\varepsilon > 0$, let $A$ be a finite subset of $J_0$ such that $|a - \sum_{i \in B} a_i| < \varepsilon$ for every finite subset $B$ of $J_0$ such that $A \subseteq B$. Note that $A \subseteq J$. Let $B'$ be a finite subset of $J$ containing $A$ and $B = B' \cap J_0$. Clearly, $\sum_{i \in B'} a_i = \sum_{i \in B} a_i$. Hence $|a - \sum_{i \in B'} a_i| < \varepsilon$, which proves the desired result. $\qquad\square$

The result of the following lemma generalizes arguments in Example 7.6.

**Lemma 7.6.** *Let $\{e_i\}_{j \in J}$ be an orthonormal family in an inner product space $X$. For every nonempty finite subset $A$ of $J$ and $x \in X$, we have*

$$\left(x - \sum_{i \in A}\langle x, e_i\rangle e_i\right) \perp e_k, \qquad \text{for all } k \in A.$$

*In particular, the vector $x - \sum_{i \in A}\langle x, e_i\rangle e_i$ is orthogonal to $\sum_{i \in A}\langle x, e_i\rangle e_i$.*

*Proof.* We have

$$\left\langle x - \sum_{i \in A}\langle x, e_i\rangle e_i, e_k\right\rangle = \langle x, e_k\rangle - \langle x, e_k\rangle = 0,$$

because $\{e_i\}_{i \in A}$ is an orthonormal family. The result follows.  □

**Theorem 7.9. (Bessel's inequality.)** *Let $\{e_i\}_{i \in J}$ be an orthonormal family of vectors in an inner product space $X$. Then for every $x \in X$, one has*

$$\sum_{i \in J}|\langle x, e_i\rangle|^2 \le \|x\|^2. \tag{7.12}$$

*Proof.* Let $A$ be a finite subset of $J$. Consider the vector

$$y = \sum_{i \in A}\langle x, e_i\rangle e_i$$

and let $z = x - y$. By Lemma 7.6, $z \perp y$. By applying the Pythagorean theorem (cf. Exercise 7.21), we obtain

$$\|x\|^2 = \|z + y\|^2 = \|z\|^2 + \|y\|^2 = \|z\|^2 + \sum_{i \in A}|\langle x, e_i\rangle|^2.$$

Because $\|z\|^2 \ge 0$, we have

$$\sum_{i \in A}|\langle x, e_i\rangle|^2 \le \|x\|^2.$$

The result follows from Lemma 7.2 (with $M = \|x\|^2$).  □

**Definition 7.7.** The inner products $\langle x, e_i\rangle$ in (7.12) are called the *Fourier coefficients* of $x$ with respect to the orthonormal family $\{e_i\}_{i \in J}$.

Note that by Lemma 7.4, there is at most a countable number of Fourier coefficients of a vector with respect to an orthonormal family in an inner product space.

*Example 7.12.* For the family $\{e_k\}_{k \in \mathbb{N}}$ of vectors in the Schauder basis for the space $\ell_2$, we have equality in (7.12) (cf. Example 7.9). However, for $x = e_1$ and the orthonormal set $\{e_2, e_3, \ldots\}$, we have strict inequality:

$$\sum_{k=2}^{\infty} |\langle e_1, e_k \rangle|^2 = 0 < \|e_1\|^2 = 1.$$

The strict Bessel's inequality in Example 7.12 is not surprising. Indeed, if $x$ is a nonzero vector in an inner product space that is orthogonal to the orthonormal family $\{e_i\}_{j \in J}$, then the left-hand side in (7.12) is zero, whereas $\|x\| \neq 0$. In this case, one may say that $\{e_i\}_{j \in J}$ is not "big enough" to produce equality in Bessel's inequality. The following definition addresses this problem.

**Definition 7.8.** An orthonormal family $\{e_i\}_{j \in J}$ in an inner product space $X$ is said to be *total* if its orthogonal complement in $X$ is the zero vector.

**Lemma 7.7.** *If $E = \{e_i : i \in J\}$ is a total orthonormal set in a Hilbert space $H$, then $\overline{\text{span}\, E} = H$.*

*Proof.* Suppose that $\overline{\text{span}\, E}$ is a proper closed subspace of $H$. Then by Theorem 7.8, the orthogonal complement of this subspace contains a nonzero vector, which contradicts the totality of the set $E$. Hence $\overline{\text{span}\, E} = H$.  $\square$

**Theorem 7.10.** *For a total orthonormal family $\{e_i\}_{i \in J}$ in a Hilbert space $H$ and a vector $x \in H$, the family $\{\langle x, e_i \rangle e_i\}_{i \in J}$ is summable with sum $x$, that is,*

$$\sum_{i \in J} \langle x, e_i \rangle e_i = x. \tag{7.13}$$

*Proof.* By Lemma 7.7, for $\varepsilon > 0$, there exist a finite set $A \subseteq J$ and a family of scalars $\{\lambda_i\}_{i \in A}$ such that

$$\left\| x - \sum_{i \in A} \lambda_i e_i \right\| < \varepsilon.$$

For every finite subset $B$ of $J$ such that $A \subseteq B$, we have

$$\left\| x - \sum_{i \in A} \lambda_i e_i \right\|^2 = \left\| x - \sum_{i \in B} \langle x, e_i \rangle e_i + \sum_{i \in B} \langle x, e_i \rangle e_i - \sum_{i \in A} \lambda_i e_i \right\|^2.$$

By Lemma 7.6, the vectors

$$x - \sum_{i \in B} \langle x, e_i \rangle e_i \qquad \text{and} \qquad \sum_{i \in B} \langle x, e_i \rangle e_i - \sum_{i \in A} \lambda_i e_i$$

are orthogonal. (Note that $A \subseteq B$.) Therefore, by the Pythagorean theorem,

$$\left\| x - \sum_{i \in A} \lambda_i e_i \right\|^2 = \left\| x - \sum_{i \in B} \langle x, e_i \rangle e_i \right\|^2 + \left\| \sum_{i \in B} \langle x, e_i \rangle e_i - \sum_{i \in A} \lambda_i e_i \right\|^2.$$

Hence,

$$\left\| x - \sum_{i \in B} \langle x, e_i \rangle e_i \right\| \leq \left\| x - \sum_{i \in A} \lambda_i e_i \right\| < \varepsilon,$$

and the result follows (cf. Definition 7.6). $\qquad\qquad\qquad\qquad\qquad\qquad\square$

**Definition 7.9.** A total orthonormal family $\{e_i\}_{i \in J}$ in a Hilbert space $H$ is called a *Hilbert basis* of $H$.

Note that the numbers $\langle x, e_i \rangle$ in (7.13) are the Fourier coefficients of $x$ with respect to the Hilbert basis $\{e_i\}_{i \in J}$ (cf. Definition 7.7).

**Theorem 7.11. (Parseval's identity.)** *An orthonormal family $\{e_i\}_{i \in J}$ in a Hilbert space $H$ is a Hilbert basis of $H$ if and only if the family $\{|\langle x, e_i \rangle e_i|^2\}_{i \in J}$ is summable for every $x \in H$, and*

$$\sum_{i \in J} |\langle x, e_i \rangle|^2 = \|x\|^2. \tag{7.14}$$

*Proof.* (Necessity.) By Theorem 7.10, $x = \sum_{i \in J} \langle x, e_i \rangle e_i$. Therefore, for every $\varepsilon > 0$, there exists a finite set $A \subseteq J$ such that

$$\left\| x - \sum_{i \in B} \langle x, e_i \rangle e_i \right\| < \sqrt{\varepsilon},$$

for every finite set $B$ such that $A \subseteq B \subseteq J$. Note that by the Pythagorean theorem,

$$\left\| \sum_{i \in B} \langle x, e_i \rangle e_i \right\|^2 = \sum_{i \in B} |\langle x, e_i \rangle|^2.$$

By Lemma 7.6, the vectors $x - \sum_{i \in B} \langle x, e_i \rangle e_i$ and $\sum_{i \in B} \langle x, e_i \rangle e_i$ are orthogonal. Hence

$$\|x\|^2 = \left\| x - \sum_{i \in B} \langle x, e_i \rangle e_i \right\|^2 + \sum_{i \in B} |\langle x, e_i \rangle|^2.$$

It follows that

$$\left| \|x\|^2 - \sum_{i \in B} |\langle x, e_i \rangle|^2 \right| < \varepsilon,$$

for every finite subset $B$ of $J$ such that $A \subseteq B$. Thus $\{|\langle x, e_i \rangle|^2\}_{i \in J}$ is summable with sum $\|x\|^2$ (cf. Definition 7.6).

(Sufficiency.) Suppose that (7.14) holds for every $x \in H$, but the family $\{e_i\}_{i \in J}$ is not total. Then there exists a nonzero vector $x \in H$ that is orthogonal to every vector in the family $\{e_i\}_{i \in J}$. For this vector $x$, the left-hand side of (7.14) is zero, whereas the right-hand side is $\|x\|^2 \neq 0$. This contradiction proves that $\{e_i\}_{i \in J}$ is a Hilbert basis of $H$. $\qquad\square$

**Theorem 7.12.** *Every Hilbert space has a Hilbert basis.*

*Proof.* Let $\mathcal{D}$ be the family of all orthonormal sets of $H$, partially ordered by inclusion. This family is not empty, because for a nonzero vector $x \in H$, the vector $e = x/\|x\|$ trivially forms an orthonormal set $\{e\}$ in $H$. Every chain in $\mathcal{D}$ has a maximal element, namely the union of all sets in the chain. By Zorn's lemma, $\mathcal{D}$ has a maximal orthonormal set $E = \{e_i : i \in J\}$. We prove that $E$ is a total set. If not, then there exists a nonzero vector $y \in H$ that is orthogonal to all vectors in $E$. Clearly, the set $E \cup \{y/\|y\|\}$ is an orthonormal set different from $E$ and containing $E$. This contradicts the maximality of $E$. The result follows. $\qquad\square$

The "canonical" example of a Hilbert space is the space $\ell_2$, which has a countable Hilbert basis (cf. Example 7.10). Recall that the finite-dimensional space $\ell_2^n$ is the $n$-dimensional Euclidean space. The following example is an "uncountable" version of the space $\ell_2$.

*Example 7.13.* Let $J$ be an uncountable set and $\ell_2(J)$ the set of all functions $f : J \to \mathbf{F}$ such that the family $\{|f(i)|^2\}_{i \in J}$ is summable. For $f, g \in \ell_2(J)$, the inner product is defined by

$$\langle f, g \rangle = \sum_{i \in J} f(i)\overline{g(i)}.$$

This construction is justified in Appendix A.2.

Here is another example of an inner product space with an uncountable orthonormal family of vectors.

*Example 7.14.* Let $X$ be the vector space of all linear combinations of functions in the family $\{e^{i\lambda t}\}_{\lambda \in \mathbf{R}}$ defined on $\mathbf{R}$. We define (for $A > 0$)

$$\langle f, g \rangle = \lim_{A \to +\infty} \frac{1}{2A} \int_{-A}^{A} f(t)\overline{g(t)}\, dt, \qquad \text{for } f, g \in X.$$

It can be shown that the family $\{e^{i\lambda t}\}_{\lambda \in \mathbf{R}}$ is a total orthonormal family in the inner product space $X$ (cf. Exercise 7.29).

Let $H$ be a separable Hilbert space and $S$ a countable dense subset of $H$. By Theorem 7.12, there exists a basis $\{e_i\}_{i \in J}$ in $H$. Consider the family $\{B(e_i, \sqrt{2}/2)\}_{i \in J}$ of open balls in $H$. It is clear that the balls in this family are disjoint. Each of these balls contains a vector from $S$, and all these vectors are distinct. Inasmuch as $S$ is a countable set, the set $J$ is at most countable. We have established the following result.

**Theorem 7.13.** *Every basis of a separable Hilbert space is at most countable.*

We used Zorn's lemma in our proof that every Hilbert space has a basis (cf. Theorem 7.12). In the case of separable Hilbert spaces, this result can be established using the "Gram–Schmidt process," which is also an important tool in other applications.

**Theorem 7.14. (Gram–Schmidt process.)** *Let $\{x_1,\ldots,x_n\}$ be a set of linearly independent vectors in an inner product space $X$. Then there exists an orthonormal set $\{e_1,\ldots,e_n\}$ such that for every $1 \le k \le n$,*

$$\text{span}\{x_1,\ldots,x_k\} = \text{span}\{e_1,\ldots,e_k\}. \tag{7.15}$$

*Furthermore, if $\{x_1,x_2,\ldots\}$ is a countable set of linearly independent vectors in $X$, then there exists a countable orthonormal set $\{e_1,e_2,\ldots\}$ in $X$ such that (7.15) holds for every $k \in \mathbf{N}$ and*

$$\text{span}\{x_1,x_2,\ldots\} = \text{span}\{e_1,e_2,\ldots\}.$$

*Proof.* The proof of the first claim is by induction on $k$. For $k = 1$, we set $e_1 = x_1/\|x_1\|$. Equality (7.15) holds trivially.

Suppose that there is an orthonormal set $\{e_1,\ldots,e_k\}$ such that (7.15) holds. By Lemma 7.6, the vector

$$y = x_{k+1} - \sum_{i=1}^{k} \langle x_{k+1}, e_i \rangle e_i$$

is orthogonal to all vectors in the set $\{e_1,\ldots,e_k\}$. We define $e_{k+1} = y/\|y\|$. It is not difficult to verify that

$$x_{k+1} \in \text{span}\{e_1,\ldots,e_k,e_{k+1}\}$$

and

$$e_{k+1} \in \text{span}\{x_1,\ldots,x_k,x_{k+1}\}.$$

Therefore,

$$\text{span}\{x_1,\ldots,x_{k+1}\} = \text{span}\{e_1,e_2,\ldots,e_{k+1}\},$$

and the result follows by induction.

To prove the second claim of the theorem, it suffices to note that

$$\text{span}\{x_1,x_2,\ldots\} = \bigcup_{k=1}^{\infty} \text{span}\{x_1,\ldots,x_k\}$$

and

$$\text{span}\{e_1,e_2,\ldots\} = \bigcup_{k=1}^{\infty} \text{span}\{e_1,\ldots,e_k\}$$

and apply (7.15).                                                              □

We apply the Gram–Schmidt process to construct a basis of a separable Hilbert space from a linearly independent set.

Let $H$ be a separable Hilbert space and $S$ a countable dense subset of $H$. Using elementary linear algebra (cf. Exercise 7.31), one can show that $S$ contains a linearly independent subset $L$ such that $\operatorname{span} L = \operatorname{span} S$. By Theorem 7.14, there is an at most countable orthonormal subset $E$ of $H$ such that $\operatorname{span} E = \operatorname{span} L$. We have

$$H = \overline{S} = \overline{\operatorname{span} S} = \overline{\operatorname{span} L} = \overline{\operatorname{span} E},$$

which means that $E$ is an orthonormal basis of $H$.

We conclude this section by establishing an important result about separable Hilbert spaces.

**Theorem 7.15.** *Every separable infinite-dimensional Hilbert space is isomorphic to $\ell_2$.*

*Proof.* Let $H$ be a separable infinite-dimensional Hilbert space. Choose an orthonormal basis $\{e'_k\}_{k \in \mathbf{N}}$ of $H$ and define an operator $T : H \to \ell_2$ by

$$x = \sum_{k=1}^{\infty} \langle x, e'_k \rangle e'_k \mapsto (\langle x, e'_1 \rangle, \ldots, \langle x, e'_k \rangle \ldots).$$

Clearly, $T$ is a linear mapping. By Parseval's identity (cf. (7.14)), $T$ is an isometry from $H$ onto a complete subspace $L$ (cf. Exercise 2.46) of $\ell_2$. Inasmuch as $Te'_k = e_k$ for all $k \in \mathbf{N}$, the subspace $L$ contains all vectors of the orthonormal basis of $\ell_2$. Hence $L^\perp = \{0\}$, which implies $T(H) = L = \ell_2$. It follows that $T$ is an isomorphism.                                                      □

**Corollary 7.3.** *All separable infinite-dimensional Hilbert spaces over the same field are isomorphic.*

## 7.4 Linear Functionals on Hilbert Spaces

As was shown in Section 5.1 (cf. Corollary 5.1), the dual of the normed space $\ell_2$ is isomorphic to the space $\ell_2$ itself. In this section we establish a similar result for Hilbert spaces.

**Theorem 7.16. (Riesz representation theorem.)** *Let $H$ be a Hilbert space and $H^*$ its dual. For every $f \in H^*$, there exists a unique $y \in H$ such that*

$$f(x) = \langle x, y \rangle, \qquad \text{for every } x \in H. \tag{7.16}$$

*Proof.* For $f \in H^*$, let $H_1$ be the null space $\mathcal{N}(f)$ of $f$. If $H_1 = H$, then $f$ is the zero functional, and we may choose $y = 0$ in (7.16).

Otherwise, let $H_2 = H_1^{\perp}$. Because $H_1 = \mathcal{N}(f)$ is a closed subspace of $H$ (cf. Corollary 4.1), we have, by Theorem 7.8, $H = H_1 \oplus H_2$. Let $y_0$ be a vector in the subspace $H_2$ such that $\|y_0\| = 1$. For every vector $x \in H$, we have

$$\langle x, y_0 \rangle = \left\langle x - \frac{f(x)}{f(y_0)} y_0 + \frac{f(x)}{f(y_0)} y_0, y_0 \right\rangle$$

$$= \left\langle x - \frac{f(x)}{f(y_0)} y_0, y_0 \right\rangle + \frac{f(x)}{f(y_0)} \langle y_0, y_0 \rangle = \frac{f(x)}{f(y_0)},$$

because

$$x - \frac{f(x)}{f(y_0)} y_0 \in \mathcal{N}(f) = H_1 \qquad \text{and} \qquad y_0 \perp H_1.$$

Let $y = \overline{f(y_0)}\, y_0$. Then $\langle x, y \rangle = f(x)$ for all $x \in H$.

To prove uniqueness of the representation, suppose that

$$f(x) = \langle x, y_1 \rangle = \langle x, y_2 \rangle, \qquad \text{for all } x \in H.$$

Then $\langle x, y_1 - y_2 \rangle = 0$ for all $x \in H$, which implies $(y_1 - y_2) \in H^{\perp} = \{0\}$. Hence $y_1 = y_2$. $\qquad\square$

By the Riesz representation theorem, every linear functional on a Hilbert space $H$ is of the form $f_y(x) = \langle x, y \rangle$, where $y$ is a unique vector in $H$, so there is a one-to-one correspondence $T$ between elements of $H$ and its dual space $H^*$.

**Theorem 7.17.** *The correspondence $T : H \to H^*$ defined by $y \mapsto f_y$, $y \in H$, is a conjugate-linear isometry.*

A general remark is in order. For vector spaces $X$ and $Y$ over $\mathbf{F}$, an operator $T : X \to Y$ is said to be *conjugate-linear* if

$$T(\lambda u + \mu v) = \overline{\lambda} T u + \overline{\mu} T v$$

for all $u \in X$, $v \in Y$, and $\lambda, \mu \in \mathbf{F}$. If $\mathbf{F}$ is the field of real numbers, then a conjugate-linear operator is just a linear one.

*Proof.* For $u, v \in H$ and $\lambda, \mu \in \mathbf{F}$, we have

$$T(\lambda u + \mu v)(x) = \langle x, \lambda u + \mu v \rangle = \overline{\lambda}\langle x, u \rangle + \overline{\mu}\langle x, v \rangle = \overline{\lambda} T u + \overline{\mu} T v.$$

Therefore, $T$ is a conjugate-linear map.

Furthermore, by the Cauchy–Schwarz inequality (cf. (7.5)),

$$|f_y(x)| = |\langle x, y \rangle| \le \|x\| \|y\|,$$

implying $\|f_y\| \le \|y\|$. On the other hand,

$$\|y\|^2 = \langle y, y \rangle = f_y(y) \le \|f_y\|\|y\|,$$

so $\|f_y\| \ge \|y\|$. Hence $\|f_y\| = \|y\|$. It follows that $T$ is an isometry.  □

**Theorem 7.18.** *The space $H^*$ is a Hilbert space with the inner product defined by*

$$\langle Tx, Ty \rangle_1 = \langle y, x \rangle, \qquad for\ all\ x, y \in H, \tag{7.17}$$

*where $T$ is as in Theorem 7.17. The space $H$ is reflexive.*

*Proof.* It can be readily verified (cf. Exercise 7.33) that (7.17) defines an inner product on $H^*$ that induces a norm on $H^*$. As a dual space, $H^*$ is complete and therefore is a Hilbert space.

To establish the reflexivity of $H$, we need to show that the canonical mapping $C : H \to H^{**}$ (cf. Definition 5.2) is surjective. For $g \in H^{**}$, we have, by the Riesz representation theorem,

$$g(f_y) = \langle Ty, Tx \rangle_1 = \langle x, y \rangle = f_y(x), \qquad for\ some\ x \in H.$$

Hence $Cx = g$, and the result follows because $g$ was an arbitrary element of the space $H^{**}$.  □

## 7.5 Sesquilinear Forms

According to the Riesz representation theorem, bounded linear functionals on a Hilbert space are completely determined by the inner product. In a similar way, bounded linear operators on Hilbert spaces are determined by two-variable functions called "sesquilinear forms."

**Definition 7.10.** Let $X$ and $Y$ be vector spaces over the field $\mathbf{F}$. A *sesquilinear form* is a function $S : X \times Y \to \mathbf{F}$ such that for all $x, x_1, x_2 \in X$, $y, y_1, y_2 \in Y$, and $\lambda_1, \lambda_2 \in \mathbf{F}$, we have

$$S(\lambda_1 x_1 + \lambda_2 x_2, y) = \lambda_1 S(x_1, y) + \lambda_2 S(x_2, y)$$

and

$$S(x, \lambda_1 y_1 + \lambda_2 y_2) = \overline{\lambda}_1 S(x, y_1) + \overline{\lambda}_2 S(x, y_2).$$

Thus the sesquilinear form $S$ is linear in the first variable and conjugate-linear in the second variable. This is why this function is called sesquilinear, which means "one and a half linear."

If $X$ and $Y$ in Definition 7.10 are normed spaces and there exists a real number $C$ such that

$$|S(x,y)| \leq C\|x\|\|y\|, \qquad \text{for all } x \in X \text{ and } y \in Y,$$

then $S$ is said to be *bounded*. In this case, the number

$$\|S\| = \sup\{|S(x,y)| : \|x\| = 1, \|y\| = 1\}$$

is called the *norm* of $S$. (The reader may want to compare these definitions with definitions of operator and functional norms in Section 4.2.)

Clearly, the inner product on an inner product space $X$ is a sesquilinear form on $X \times X$. By the Cauchy–Schwarz inequality, this form is bounded.

For a bounded sesquilinear form $S$, we have

$$|S(x,y)| \leq \|S\|\|x\|\|y\|$$

(cf. Exercise 7.34).

**Theorem 7.19.** *Let $H_1$ and $H_2$ be Hilbert spaces and*

$$S : H_1 \times H_2 \to \mathbf{F}$$

*a bounded sesquilinear form. Then there exists a bounded linear operator $T$ from $H_1$ into $H_2$ such that*

$$S(x,y) = \langle Tx, y \rangle, \tag{7.18}$$

*for all $x \in X$ and $y \in Y$. This operator is uniquely determined by $S$ and has the same norm as $S$:*

$$\|T\| = \|S\|. \tag{7.19}$$

*Proof.* For a fixed $x \in H_1$, the function $\overline{S(x,y)}$ is a linear functional on $H_2$. Therefore, by the Riesz representation theorem, there exists $z \in H_2$ such that

$$\overline{S(x,y)} = \langle y, z \rangle,$$

which yields

$$S(x,y) = \overline{\langle y, z \rangle} = \langle z, y \rangle,$$

where $z$ is unique but depends on the fixed $x \in H_1$. We define $Tx = z$, which immediately implies (7.18). It remains to show that $T$ is the desired operator.

First, we show that $T$ is a linear operator. Indeed,

$$\langle T(\lambda_1 x_1 + \lambda_2 x_2), y \rangle = S(\lambda_1 x_1 + \lambda_2 x_2, y) = \lambda_1 S(x_1, y) + \lambda_2 S(x_2, y)$$
$$= \lambda_1 \langle Tx_1, y \rangle + \lambda_2 \langle Tx_2, y \rangle = \langle \lambda_1 Tx_1 + \lambda_2 Tx_2, y \rangle,$$

for all $y \in H_2$. By Exercise 7.32, we obtain

$$T(\lambda_1 x_1 + \lambda_2 x_2) = \lambda_1 T x_1 + \lambda_2 T x_2.$$

Second, we show that $T$ is bounded. We may assume that $T \neq 0$. In this case, there exists $x \in X$ such that $Tx \neq 0$. We have

$$
\begin{aligned}
\|S\| &= \sup\{|S(x,y)| : \|x\| = 1, \|y\| = 1\} \\
&= \sup\{|\langle Tx, y \rangle| : \|x\| = 1, \|y\| = 1\} \\
&\geq \sup\{|\langle Tx, Tx/\|Tx\| \rangle| : \|x\| = 1, Tx \neq 0\} \\
&= \sup\{\|Tx\| : \|x\| = 1\} = \|T\|.
\end{aligned}
$$

(Note that $\langle Tx, Tx/\|Tx\| \rangle = \|Tx\|$.)

Finally, we prove (7.19). By applying (7.18) and the Cauchy–Schwarz inequality, we obtain

$$
\begin{aligned}
\|S\| &= \sup\{|\langle Tx, y \rangle| : \|x\| = 1, \|y\| = 1\} \\
&\leq \sup\{\|T\|\|x\|\|y\| : \|x\| = 1, \|y\| = 1\} = \|T\|.
\end{aligned}
$$

The last two displayed inequalities yield $\|T\| = \|S\|$.  □

## 7.6 Hilbert-Adjoint Operator

The notion of the adjoint operator plays an important role in linear algebra, where it is used to describe important classes of operators on inner product spaces, especially, in connection with the spectral theorem. The adjoint of a linear operator $T : X \to Y$, where $X$ and $Y$ are inner product spaces, is the operator $T^* : Y \to X$ such that

$$\langle Tx, y \rangle = \langle x, T^* y \rangle,$$

for every $x \in X$ and every $y \in Y$.

*Example 7.15.* The inner product on the Euclidean space $\ell_2^n$ is defined by

$$\langle x, y \rangle = \sum_{k=1}^{n} x_k \overline{y_k}$$

(cf. (7.1)). It is known from linear algebra that the adjoint of a linear map defined by a matrix $A$ is given by the conjugate transpose matrix $A^* = \overline{A^t}$. Note that $A^* = A^t$ in the real case.

Consider now a linear operator $T : H_1 \to H_2$, where $H_1$ and $H_2$ are Hilbert spaces. If we want to extend the definition of adjoint to operators on Hilbert spaces, it is natural to look for an operator $S : H_2 \to H_1$ such that

$$\langle Tx, y \rangle = \langle x, Sy \rangle, \tag{7.20}$$

for all $x \in H_1$ and $y \in H_2$. The following theorem shows that the algebraic condition (7.20) is strong enough to imply the continuity of operators $T$ and $S$.

**Theorem 7.20.** *Let $H_1$ and $H_2$ be Hilbert spaces. If the operators*

$$T : H_1 \to H_2 \qquad and \qquad S : H_2 \to H_1$$

*satisfy the equation*

$$\langle Tx, y \rangle = \langle x, Sy \rangle$$

*for all $x \in H_1$ and $y \in H_2$, then $T$ and $S$ are bounded.*

*Proof.* We prove that $T$ is bounded by applying the closed graph theorem (cf. Theorem 6.6). The proof for the operator $S$ is similar.

Let $(x_n)$ and $(Tx_n)$ be sequences such that $x_n \to x$ and $Tx_n \to y$ in $H_1$ and $H_2$, respectively. For every $z \in H_2$, we have

$$\langle y, z \rangle = \lim_{n \to \infty} \langle Tx_n, z \rangle = \lim_{n \to \infty} \langle x_n, Sz \rangle = \langle x, Sz \rangle = \langle Tx, z \rangle,$$

because inner products are continuous functions. It follows that $Tx = y$. By Theorem 6.6, $T$ is a bounded operator. $\qquad\qquad\square$

The following corollary is known as the *Hellinger–Toeplitz theorem.* An operator $T$ on a Hilbert space $H$ is said to be *symmetric* if $\langle Tx, y \rangle = \langle x, Ty \rangle$ for all $x, y \in H$.

**Corollary 7.4.** *A symmetric operator on a Hilbert space is bounded.*

The result of Theorem 7.20 motivates the following definition.

**Definition 7.11.** For Hilbert spaces $H_1$ and $H_2$ and a bounded operator $T : H_1 \to H_2$, the *Hilbert-adjoint operator $T^*$ of $T$* is the operator

$$T^* : H_2 \to H_1$$

such that
$$\langle Tx, y \rangle = \langle x, T^*y \rangle, \qquad \text{for all } x \in H_1, y \subset H_2. \tag{7.21}$$

(Note that the inner products on the left- and right-hand sides of the above equation are on the spaces $H_2$ and $H_1$, respectively.)

Theorem 7.21 justifies this definition.

**Theorem 7.21.** *The Hilbert-adjoint $T^*$ of $T$ (cf. Definition 7.11) exists, is unique, and is a bounded linear operator with the norm*

$$\|T^*\| = \|T\|.$$

*Proof.* The function $S(y,x) = \langle y, Tx \rangle$ on $H_2 \times H_1$ is a sesquilinear form. Indeed, it is clearly linear in the first argument and conjugate-linear in the second argument. By the Cauchy–Schwarz inequality, $S$ is a bounded form:

$$|S(y,x)| = |\langle y, Tx \rangle| \leq \|y\| \|Tx\| \leq \|T\| \|x\| \|y\|.$$

Hence $\|S\| \leq \|T\|$.

Furthermore, for $x \in H_1$ and $y \in H_2$,

$$\|S\| \geq \frac{|S(y,x)|}{\|y\| \|x\|} = \frac{|\langle y, Tx \rangle|}{\|y\| \|x\|}.$$

By taking $y = Tx$ in the above inequality, we obtain

$$\|S\| \geq \frac{|\langle Tx, Tx \rangle|}{\|Tx\| \|x\|} = \frac{\|Tx\|}{\|x\|}.$$

By the definition of $\|T\|$, we have $\|S\| \geq \|T\|$. By combining this inequality with the last inequality in the preceding paragraph, we obtain $\|S\| = \|T\|$.

By Theorem 7.19, there exists a unique operator $T^* : H_2 \to H_1$ such that

$$S(y,x) = \langle T^*y, x \rangle, \qquad \text{for all } x \in H_1 \text{ and } y \in H_2,$$

and $\|T^*\| = \|S\|$, which implies $\|T^*\| = \|T\|$. From the above displayed equality, we obtain

$$\langle y, Tx \rangle = \langle T^*y, x \rangle, \qquad \text{for all } x \in H_1 \text{ and } y \in H_2,$$

which is equivalent to

$$\langle Tx, y \rangle = \langle x, T^*y \rangle, \qquad \text{for all } x \in H_1 \text{ and } y \in H_2.$$

Therefore, $T^*$ is the Hilbert-adjoint of $T$.                                           $\square$

*Example 7.16.* Let $S$ and $T$ be the right and left shift operators on the Hilbert space $\ell_2$ defined by

$$S : (x_1, x_2, x_3, \dots) \mapsto (0, x_1, x_2, \dots)$$

and

$$T : (x_1, x_2, x_3, \dots) \mapsto (x_2, x_3, x_4, \dots)$$

(cf. Example 4.4). It is easy to verify that $S$ and $T$ are linear operators and $\|S\| = \|T\| = 1$ (cf. Exercise 4.9). We have

$$\langle x, Sy \rangle = x_2 \overline{y_1} + x_3 \overline{y_2} + \cdots = \langle Tx, y \rangle,$$

for all $x, y \in \ell_2$. Hence $T^* = S$.

There are useful properties of Hilbert-adjoint operators, some of which are known from linear algebra.

**Theorem 7.22.** *Let $H$ be a Hilbert space, $S : H \to H$ and $T : H \to H$ bounded linear operators, and $\lambda$ a scalar. Then*

(a)   $(S+T)^* = S^* + T^*$,
(b)   $(\lambda T)^* = \bar{\lambda} T^*$,
(c)   $(ST)^* = T^* S^*$,
(d)   $T^{**} = T$,
(e)   $\|T^*T\| = \|T\|^2$.

*Proof.* Parts (a) and (b) follow immediately from (7.21) and Exercise 7.32, which we also use below for parts (c) and (d) of the theorem.

For all $x, y \in H$, we have

$$\langle (ST)^* x, y \rangle = \langle x, STy \rangle = \langle S^* x, Ty \rangle = \langle T^* S^* x, y \rangle,$$

so $(ST)^* = T^* S^*$, and

$$\langle T^{**} x, y \rangle = \langle (T^*)^* x, y \rangle = \langle x, T^* y \rangle = \langle Tx, y \rangle,$$

so $T^{**} = T$.

Finally, for part (e), we have (cf. Exercise 4.11)

$$\|T^*T\| \leq \|T^*\| \|T\| = \|T\|^2,$$

and by the Cauchy–Schwarz inequality, for every $x \in H$,

$$\|Tx\|^2 = \langle Tx, Tx \rangle = \langle T^*Tx, x \rangle \leq \|T^*Tx\| \|x\| \leq \|T^*T\| \|x\|^2,$$

which implies $\|T\|^2 \leq \|T^*T\|$. Hence $\|T^*T\| = \|T\|^2$.     □

Recall that we denote by $\mathcal{N}(T)$ the null space of an operator $T : H \to H$. Similarly, we denote by $\mathcal{R}(T) = T(H)$ the *range* of $T$.

**Theorem 7.23.** *For all bounded linear operators $T : H \to H$,*

(a)   $\mathcal{N}(T^*) = \mathcal{R}(T)^{\perp}$,
(b)   $\mathcal{N}(T) = \mathcal{R}(T^*)^{\perp}$.

*Proof.* For $x \in H$, we have $x \in \mathcal{N}(T^*)$ if and only if $T^* x = 0$, which occurs if and only if $\langle T^* x, y \rangle = 0$, for all $y \in H$, which occurs if and only if $\langle x, Ty \rangle = 0$, for all $y \in H$, which occurs if and only if $x \in \mathcal{R}(T)^{\perp}$.

By replacing $T$ by $T^*$ in part (a), we prove part (b), because $T^{**} = T$.     □

## 7.7 Normal, Self-adjoint, and Unitary Operators

In this section we introduce three major classes of bounded linear operators on a Hilbert space and establish their properties. In the finite-dimensional case, these classes are well known from linear algebra.

**Definition 7.12.** A bounded linear operator $T$ on a Hilbert space $H$ is said to be

1.   *normal* if $TT^* = T^*T$,
2.   *self-adjoint* if $T^* = T$,
3.   *unitary* if $TT^* = T^*T = I$, where $I$ is the identity operator on $H$.

It clear from the definition that self-adjoint and unitary operators are normal.

*Example 7.17.* For a given $a = (a_n) \in \ell_\infty$, define an operator $T_a : \ell_2 \to \ell_2$ by

$$T_a(x_n) = (a_n x_n), \qquad \text{for } (x_n) \in \ell_2.$$

Clearly, $T_a$ is a linear operator, and

$$\|T_a(x_n)\|_2 = \sqrt{\sum_{k=1}^{\infty} |a_n x_n|^2} \leq \|(a_n)\|_\infty \|(x_n)\|_2$$

shows that $T_a$ is bounded. For all $x = (x_n)$ and $y = (y_n)$ in the Hilbert space $\ell_2$, we have

$$\langle T_a\, x, y \rangle = \sum_{k=1}^{\infty} (a_k x_k) \overline{y_k} = \sum_{k=1}^{\infty} x_k \,\overline{\overline{a_k}\, y_k} = \langle x, T_{\overline{a}}\, y \rangle,$$

where $\overline{a} = (\overline{a_n})$. Hence $T_a^* = T_{\overline{a}}$.

Because $T_a T_b = T_b T_a = T_{ab}$ for all $a, b \in \ell_\infty$, it follows that $T_a$ is normal. Note that in the real case, $T_a^* = T_a$, so $T_a$ is self-adjoint.

*Example 7.18.* Let $S$ and $T$ be the right and left shift operators on $\ell_2$ (cf. Example 7.16). We have

$$SS^* = ST : (x_1, x_2, x_3, \ldots) \mapsto (0, x_2, x_3, \ldots)$$

and

$$S^*S = TS : (x_1, x_2, x_3, \ldots) \mapsto (x_1, x_2, x_3, \ldots).$$

It follows that $S$ and $T$ are not normal operators.

In the following example we present a sample of operators that illustrate relations between classes of normal, self-adjoint, and unitary operators. For simplicity's sake, we consider the Hilbert space $\ell_2^2$ with a standard basis, and represent operators by their matrices with respect to this basis. The reader should verify the statements in this example (cf. Exercise 7.41).

*Example 7.19.*

(A) $\quad \begin{pmatrix} 2i & 0 \\ 0 & 2i \end{pmatrix} \qquad$ normal, not self-adjoint, not unitary

(B) $\quad \begin{pmatrix} 0 & 2 \\ 2 & 0 \end{pmatrix} \qquad$ self-adjoint, not unitary

(C) $\quad \begin{pmatrix} \cos\varphi & -\sin\varphi \\ \sin\varphi & \cos\varphi \end{pmatrix} \qquad 0 < \varphi < \pi/2 \quad$ unitary not self-adjoint

(D) $\quad \begin{pmatrix} 0 & 1 \\ 1 & 0 \end{pmatrix} \qquad$ unitary and self-adjoint

The three major classes of bounded operators are shown schematically in Fig. 7.3.

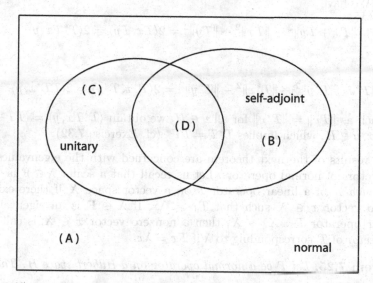

**Fig. 7.3** Classes of bounded operators on Hilbert space.

In the following two theorems we establish some properties of normal operators.

**Theorem 7.24.** *An operator $T$ on a Hilbert space $H$ is normal if and only if $\|Tx\| = \|T^*x\|$ for all $x \in H$.*

*Proof.* (Necessity.) Suppose that $T : H \to H$ is a normal operator, that is, $TT^* = T^*T$. We have

$$\|Tx\|^2 = \langle Tx, Tx \rangle = \langle T^*Tx, x \rangle = \langle TT^*x, x \rangle = \langle T^*x, T^*x \rangle = \|T^*x\|^2,$$

so $\|Tx\| = \|T^*x\|$ for all $x \in H$.

(Sufficiency.) First, suppose that $\|Tx\| = \|T^*x\|$ for all $x$ in a complex Hilbert space $H$. Then

$$\|Tx\|^2 = \langle Tx, Tx \rangle = \langle T^*Tx, x \rangle$$

and

$$\|T^*x\|^2 = \langle T^*x, T^*x \rangle = \langle TT^*x, x \rangle.$$

Therefore, $\langle T^*Tx, x \rangle = \langle TT^*x, x \rangle$ for all $x \in H$, so $\langle (T^*T - TT^*)x, x \rangle = 0$, for all $x \in H$. By the result of Exercise 7.36, $T^*T = TT^*$; hence $T$ is normal.

Suppose now that $H$ is a real Hilbert space and $\|Tv\| = \|T^*v\|$ for all $v \in H$. We have

$$\|Tx + Ty\|^2 - \|Tx\|^2 - \|Ty\|^2 = 2\langle Tx, Ty \rangle = 2\langle T^*Tx, y \rangle$$

and

$$\|T^*x + T^*y\|^2 - \|T^*x\|^2 - \|T^*y\|^2 = 2\langle T^*x, T^*y \rangle = 2\langle TT^*x, y \rangle.$$

Inasmuch as $\|Tv\| = \|T^*v\|$ for all $v \in H$, we obtain $\langle T^*Tx, y \rangle = \langle TT^*x, y \rangle$, for all $x, y \in H$, which implies $T^*T = TT^*$ (cf. Exercise 7.32). $\qquad\square$

The results of the next theorem are concerned with the eigenvalues and eigenvectors of normal operators. Let us recall that a scalar $\lambda \in \mathbf{F}$ is called an *eigenvalue* of a linear operator $T$ on a vector space $X$ if there exists a nonzero vector $x \in X$ such that $Tx = \lambda x$. If $\lambda \in \mathbf{F}$ is an eigenvalue of a linear operator $T : X \to X$, then a nonzero vector $x \in X$ is called an *eigenvector* of $T$ corresponding to $\lambda$ if $Tx = \lambda x$.

**Theorem 7.25.** *Let $T$ be a normal operator on a Hilbert space $H$. Then:*

(a) *If $x \in H$ is an eigenvector of $T$ with eigenvalue $\lambda$, then $x$ is also an eigenvector of $T^*$ with eigenvalue $\overline{\lambda}$.*

(b) *Eigenvectors corresponding to distinct eigenvalues are orthogonal.*

*Proof.* (a) The operator $T - \lambda I$ is normal (cf. Exercise 7.43). Therefore, by Exercise 7.42 and Theorem 7.24,

$$0 = \|(T - \lambda I)x\| = \|(T^* - \overline{\lambda}I)x\|.$$

It follows that $x$ is an eigenvector of $T^*$ corresponding to the eigenvalue $\overline{\lambda}$.

(b) Let $x$ and $y$ be eigenvectors of $T$ corresponding to distinct eigenvalues $\lambda$ and $\mu$, so $Tx = \lambda x$, $Ty = \mu y$, and $\lambda \neq \mu$. By part (a) of the theorem,

$$\lambda \langle x, y \rangle = \langle \lambda x, y \rangle = \langle Tx, y \rangle = \langle x, T^* y \rangle = \langle x, \overline{\mu} y \rangle = \mu \langle x, y \rangle.$$

Because $\lambda \neq \mu$, it follows that $\langle x, y \rangle = 0$. Hence $x \perp y$. □

Because self-adjoint operators are normal, they have properties that were established in the previous two theorems. Additional properties that are specific to self-adjoint operators are the subject of the next two theorems.

**Theorem 7.26.** *Let $T$ be a bounded operator on a Hilbert space $H$. Then:*

(a) *If $T$ is self-adjoint, then $\langle Tx, x \rangle$ is real for all $x \in H$.*
(b) *If the space $H$ is complex and $\langle Tx, x \rangle$ is real for all $x \in H$, then $T$ is self-adjoint.*
(c) *If $T$ is self-adjoint, then all eigenvalues of $T$ are real.*

*Proof.* (a) For every $x \in H$, we have

$$\langle Tx, x \rangle = \langle x, Tx \rangle = \overline{\langle Tx, x \rangle}.$$

Hence $\langle Tx, x \rangle$ is real.

(b) If $\langle Tx, x \rangle$ is real for all $x \in H$, then

$$\langle Tx, x \rangle = \overline{\langle Tx, x \rangle} = \overline{\langle x, T^* x \rangle} = \langle T^* x, x \rangle,$$

for all $x \in H$. Because $H$ is complex, $T^* = T$ (cf. Exercise 7.36).

(c) Suppose that $Tx = \lambda x$ for some $x \neq 0$. By Theorem 7.25(a), $Tx = \overline{\lambda} x$. Hence, $\lambda x = \overline{\lambda} x$, which implies $\lambda = \overline{\lambda}$, because $x \neq 0$. □

**Theorem 7.27.** *The product of two self-adjoint operators $S$ and $T$ on a Hilbert space is self-adjoint if and only if the operators commute, that is, $ST = TS$.*

*Proof.* By Theorem 7.22(c),

$$(ST)^* = T^* S^* = TS,$$

because $S$ and $T$ are self-adjoint. Therefore,

$$(ST)^* = ST \qquad \text{if and only if} \qquad ST = TS,$$

whence the result. □

Traditionally, unitary operators on a Hilbert space are denoted by $U$. In what follows, we discuss some properties of unitary operators. First, we establish equivalent descriptions of the unitary operators.

**Theorem 7.28.** *For a bounded linear operator $U$ on a Hilbert space $H$, the following statements are equivalent:*

(a) $U$ *is a unitary operator.*
(b) $U$ *is a bijection and* $\langle Ux, Uy \rangle = \langle x, y \rangle$ *for all* $x, y \in H$.
(c) $U$ *is a bijection and* $\|Ux\| = \|x\|$ *for every* $x \in H$.

*Proof.* (a) implies (b). For every $x \in H$, $x = UU^*x = U(U^*x)$. Hence $U$ maps $H$ onto $H$. Furthermore,

$$\langle Ux, Uy \rangle = \langle x, U^*Ux \rangle = \langle x, y \rangle, \qquad \text{for all } x, y \in H,$$

because $U^*U = I$.

(b) implies (c). Obvious.

(c) implies (a). First, suppose that $H$ is real. By Theorem 7.3 (polarization identities), for all $x, y \in H$,

$$\langle U^*Ux, y \rangle = \langle Ux, Uy \rangle = \tfrac{1}{4}\big(\|Ux + Uy\|^2 - \|Ux - Uy\|^2\big)$$
$$= \tfrac{1}{4}\big(\|x + y\|^2 - \|x - y\|^2\big) = \langle x, y \rangle.$$

Hence $U^*U = I$ (cf. Exercise 7.32).

If $H$ is a complex space, then the complex part of Theorem 7.3 yields the same result: $U^*U = I$.

Because $U$ is onto and preserves norms, it is an isometry of $H$ onto itself. Hence $U$ is invertible. Because $U^*U = I$, we have $U^* = U^{-1}$, which implies $UU^* = I$. Therefore, $U$ is a unitary operator. $\qquad\square$

**Corollary 7.5.** *A unitary operator $U : H \to H$ is an isomorphism (cf. Definition 7.2) of the Hilbert space $H$ onto itself. It is called a Hilbert automorphism of $H$.*

## 7.8 Projection Operators

Now we turn our attention to projection operators on a Hilbert space $H$. Suppose that $E$ is a closed subspace of $H$. According to Theorem 7.8, there is a decomposition $H = E \oplus E^\perp$, that is, every $x \in H$ has a unique representation

$$x = u + v, \qquad u \in E, \ v \in E^\perp. \tag{7.22}$$

The vector $u$ in this representation is said to be the *projection of $x$* (cf. Theorem 7.7) onto the subspace $E$.

Because $u$ in (7.22) is unique, the operator $P : x \mapsto u$ is well defined. Clearly, it is a linear operator on $H$. The operator $P$ is called a *projection* (or more specifically, the projection of $H$ onto $E$) and sometimes denoted by $P_E$ if projections onto different subspaces are considered simultaneously. Note that $P_{\{0\}} = 0$, the zero operator, and $P_H = I$, the identity operator.

Inasmuch as $Px \in E$ for all $x \in H$, we obtain $P(Px) = P^2x = Px$ for all $x \in H$. Hence

$$P^2 = P.$$

An operator $P$ satisfying this condition is said to be *idempotent*.

We write (7.22) as

$$x = Px + (I - P)x, \qquad \text{for all } x \in H,$$

and apply the Pythagorean theorem (cf. Exercise 7.21) to obtain

$$\|x\|^2 = \|Px\|^2 + \|(I - P)x\|^2, \qquad \text{for all } x \in H.$$

It follows that $P$ and $I - P$ are bounded operators. Moreover, $\|P\| \le 1$ and $\|I - P\| \le 1$. Because $\|P(Px)\| = \|Px\|$, we conclude that $\|P\| \ge 1$. Therefore, $\|P\| = 1$ if $P \ne 0$.

**Theorem 7.29.** *A bounded operator $P : H \to H$ is a projection if and only if $P$ is self-adjoint and $P^2 = P$.*

*Proof.* (Necessity.) Suppose that a bounded operator $P$ on a Hilbert space is an orthogonal projection, and let $E = P(H)$. We have already proved that $P^2 = P$.

Let $x = u + v$ and $x' = u' + v'$, where $u, u' \in E$ and $v, v' \in E^\perp$. Then

$$\langle Px, x' \rangle = \langle u, u' + v' \rangle = \langle u, u' \rangle = \langle u + v, u' \rangle = \langle x, Px' \rangle.$$

Hence $P$ is self-adjoint.

(Sufficiency.) Suppose that $P^2 = P = P^*$, and as before, let $E = P(H)$. The subspaces $E = P(H)$ and $(I - P)(H)$ are orthogonal. Indeed,

$$\langle Px, (I - P)y \rangle = \langle x, P(I - P)y \rangle = \langle x, (P - P^2)y \rangle = 0,$$

for all $x, y \in H$. From

$$x = Px + (I - P)x, \qquad \text{for all } x \in H,$$

we obtain $(I - P)(H) = E^\perp$, so $H = E \oplus E^\perp$.

It remains to show that $E$ is a closed subspace of $H$. First, we prove that $E = \{x \in H : Px = x\}$. If $Px = x$, then clearly, $x \in E$. On the other hand, if $x \in E$, then there exists $y \in H$ such that $x = Py$. The following shows that $Px = x$:

$$Px = P^2y = Py = x.$$

Finally, we have

$$\mathcal{N}(I - P) = \{x \in H : (I - P)x = 0\} = \{x \in H : Px = x\} = E.$$

Because $I - P$ is a bounded operator, $E$ is closed.                                    □

Theorem 7.29 establishes the correspondence between projections defined in algebraic terms with closed subspaces of a Hilbert space. The next theorem provides a simple algebraic (and geometric) property of projection operators. More properties of projections are found in Exercises 7.54–7.56.

**Theorem 7.30.** *Let $E$ and $E'$ be closed subspaces of a Hilbert space $H$ and let $P = P_E$, $Q = P_{E'}$ be the corresponding projection operators. Then $PQ$ is a projection if and only if $P$ and $Q$ commute, that is, $PQ = QP$. In this case, $PQ$ is the projection onto the closed subspace $E \cap E'$.*

*Proof.* (Necessity.) Suppose $PQ$ is a projection. Then by Theorem 7.22 (c),

$$PQ = (PQ)^* = Q^*P^* = QP.$$

(Sufficiency.) If $PQ = QP$, then

$$(PQ)^* = Q^*P^* = QP = PQ.$$

Hence $PQ$ is self-adjoint. Furthermore,

$$(PQ)^2 = (PQ)(PQ) = PQPQ = PPQQ = PQ.$$

It follows that $PQ$ is a projection.

For every $x \in H$, we have

$$(PQ)x = P(Qx) = Q(Px).$$

We have $P(Qx) \in E$ and $Q(Px) \in E'$. Therefore, $(PQ)x \in E \cap E'$, so $PQ$ projects $H$ into $E \cap E'$. If $y \in E \cap E'$, then $y \in E$ and $y \in E'$, so $Py = y$ and $Qy = y$. We have

$$(PQ)y = P(Qy) = Py = y.$$

It follows that $PQ$ projects $H$ onto $E \cap E'$.                                       □


# Notes

The inner product generalizes the dot product—a concept well known from Euclidean geometry and physics. In the Euclidean space $\mathbf{R}^3$, the dot product of two vectors $\boldsymbol{x} = (x_1, x_2, x_3)$ and $\boldsymbol{y} = (y_1, y_2, y_3)$ is defined by

$$\boldsymbol{x} \cdot \boldsymbol{y} = x_1y_1 + x_2y_2 + x_3y_3,$$

and equivalently, by $\boldsymbol{x} \cdot \boldsymbol{y} = |\boldsymbol{x}||\boldsymbol{y}|\cos\vartheta$, where $|\boldsymbol{x}|$ and $|\boldsymbol{y}|$ are the respective magnitudes of the vectors $\boldsymbol{x}$ and $\boldsymbol{y}$, and $\vartheta$ is the angle between these two vectors. In physics, for instance, the mechanical work is the dot product of the force and displacement vectors.

Theorem 7.7 is an instance of a more general result that can be easily deduced from the proof of the theorem (cf. Exercise 7.23). Namely, let $E$ be a closed convex subset of a Hilbert space. Then for every $x \in H$, there is a unique $y \in E$ such that

$$\|x - y\| = d(x, E) = \inf_{u \in E} \|x - u\|.$$

According to Lemma 7.4, a family $\{a_i\}_{i \in J}$ of nonnegative numbers is summable in $\mathbf{R}$ only if the set of indices $i$ such that $a_i \neq 0$ is at most countable. Thus, the study of summable families reduces essentially to the study of summable sequences. However, as examples in Section 7.3 and Appendix A demonstrate, we have to consider possibly uncountable families in the main theorems of Section 7.3. Our definition of summable families (cf. Definition 7.6) is adopted from Bourbaki (1966), Section 5.1.

All bases of a Hilbert space $H$ have the same cardinality. This result can be easily established for separable spaces (including finite-dimensional spaces). However, for a general $H$ this is not a simple fact. The proof can be found, for instance, in Dunford and Schwartz (1958), (Chapter IV, Section 4, Theorem 14). The cardinality of a basis of the space $H$ is called the *Hilbert dimension* of $H$. It can be proved (cf. Corollary 7.3) that every two Hilbert spaces of the same dimension are isomorphic.

We did not prove that the space $\ell_2(.J)$ from Example 7.13 is indeed a Hilbert space. For this, the reader is referred to Appendix A.2.

The vector space $X$ from Example 7.14 becomes a normed space if the norm is defined by

$$\|f\| = \sup_{x \in \mathbf{R}} |f(x)|.$$

The completion of $X$ with respect to this norm is a normed space of *almost periodic* functions on $\mathbf{R}$ (Dunford and Schwartz 1958, Chapter IV, Section 7). Note that this norm is not the same as the norm generated by the inner product introduced in Example 7.14.

There are unbounded operators $T$ on a Hilbert space in quantum mechanics that satisfy the relation $\langle Tx, y \rangle = \langle x, Ty \rangle$. By the Hellinger–Toeplitz theorem (cf. Corollary 7.4), such operators cannot be defined on the entire space. The theory of unbounded operators is beyond the scope of this book.

The classes of normal, self-adjoint, unitary, and projection operators have been studied extensively because of their role in more advanced parts of functional analysis and its applications (see, for instance, Chapter VI in Dunford and Schwartz 1958), where other important classes of operators are also introduced and studied).

The conditions $P^2 = P = P^*$ are often used as the definition of the projection operator $P$. Projection operators play a significant role in the spectral theory of self-adjoint operators.

In geometry, Apollonius's identity (cf. Exercise 7.10) states that the sum of the squares of any two sides of any triangle equals twice the square of half the third side, together with twice the square of the median bisecting the third side.

## Exercises

**7.1.** Let $X$ be an inner product space. Show that

1. $\langle \lambda x + \mu y, z \rangle = \lambda \langle x, z \rangle + \mu \langle y, z \rangle$,
2. $\langle x, \lambda y + \mu z \rangle = \overline{\lambda} \langle x, y \rangle + \overline{\mu} \langle x, z \rangle$,

for every $x, y, z \in X$ and $\lambda, \mu \in \mathbf{F}$.

**7.2.** Prove that (7.1) defines an inner product on $\ell_2$.

**7.3.** Prove that the integral of a continuous nonnegative function $x(t)$ on the interval $[a, b]$ is zero if and only if the function $x$ is the zero function.

**7.4.** Let $X_1$ be the vector space $C[a, b]$ endowed with the norm $\| \cdot \|_\infty$, and $X_2$ the same vector space endowed with the norm

$$\|x\| = \sqrt{\int_a^b |x(t)|^2 \, dt}$$

(cf. Example 7.2). Show that the identity map $x \to x$ of $X_1$ onto $X_2$ is continuous.

**7.5.** Show that $\langle x, 0 \rangle = \langle 0, y \rangle = 0$ for all $x$ and $y$ in an inner product space.

**7.6.** Show that $\|\lambda x\| = |\lambda| \|x\|$ for every $\lambda \in \mathbf{F}$ and $x \in X$, where $X$ is an inner product space.

**7.7.** Show that $\langle x, y \rangle$ is an inner product on the space $H$ in the proof of Theorem 7.5.

**7.8.** Let $Y$ be a subspace of a Hilbert space $H$. Prove that

(a) $Y$ is complete if and only if it is closed in $H$.
(b) If $Y$ is finite-dimensional, then $Y$ is complete.

**7.9.** Let $x$ and $y$ be vectors in an inner product space $X$. Show that

$$\langle x, x \rangle \langle y, y \rangle \geq \frac{1}{4} \left( \langle x, y \rangle + \langle y, x \rangle \right)^2.$$

**7.10.** (**Apollonius's identity.**) Let $x$, $y$, and $z$ be vectors in an inner product space $X$. Show that

$$\|z - x\|^2 + \|z - y\|^2 = \tfrac{1}{2}\|x - y\|^2 + 2\|z - \tfrac{1}{2}(x + y)\|^2.$$

**7.11.** Let $(x_n)$ be a sequence in an inner product space. Show that the conditions $\|x_n\| \to \|x\|$ and $\langle x_n, x \rangle \to \langle x, x \rangle$ imply $x_n \to x$.

**7.12.** Show that

$$X = \left\{ x = (x_n) \in \ell_2 : \sum_{n=1}^{\infty} x_n/n = 0 \right\}$$

is a closed subspace of $\ell_2$.

**7.13.** Let $x$ and $y$ be linearly independent vectors in an inner product space such that $\|x\| = \|y\| = 1$. Show that

$$\|tx + (1 - t)y\| < 1, \qquad \text{for } 0 < t < 1.$$

**7.14.** If a vector $x$ is orthogonal to every vector in the set $\{x_1, \ldots, x_n\}$, then it is orthogonal to every linear combination of these vectors.

**7.15.** Prove that in an inner product space, $x \perp y$ if and only if

$$\|x + \lambda y\| = \|x - \lambda y\|,$$

for all scalars $\lambda \in \mathbf{F}$.

**7.16.** Show that in 1-dimensional Hilbert space, $x \perp y$ if and only if at least one of the vectors $x$ and $y$ is the zero vector.

**7.17.** Let $X$ be an inner product space and $E$ a nonempty subset of $X$. Show that

$$E^\perp = \bigcap_{x \in E} \{x\}^\perp.$$

**7.18.** (a) Prove that for every two subspaces $X_1$ and $X_2$ of a Hilbert space,

$$(X_1 + X_2)^\perp = X_1^\perp \cap L_2^\perp.$$

(b) Prove that for every two closed subspaces $X_1$ and $X_2$ of a Hilbert space,

$$(X_1 \cap X_2)^\perp = \overline{X_1^\perp + X_2^\perp}.$$

**7.19.** Let $X$ be an inner product space and $E$ a nonempty subset of $X$. Prove the following properties:

(a) If $x \perp E$, then $x \perp \overline{\text{span}E}$.

(b) If $E$ is dense in $X$ and $x \perp E$, then $x = 0$.
(c) The set $E^\perp$ is a closed subspace of $X$.

**7.20.** Show that the space $X$ in Example 7.8 is not complete and the subspace $E$ is closed in $X$. Also show that $E \oplus E^\perp = \{x \in X : x(0) = 0\}$.

**7.21. (Pythagorean theorem.)** Let $X$ be an inner product space. Prove that

(a) If $x \perp y$ in $X$, then $\|x + y\|^2 = \|x\|^2 + \|y\|^2$.
(b) If $\{x_1, \ldots, x_n\}$ is a set of mutually orthogonal vectors in $X$, then

$$\left\| \sum_{k=1}^n x_k \right\|^2 = \sum_{k=1}^n \|x_k\|^2.$$

(c) If $(x_n)$ is a sequence of mutually orthogonal vectors in $X$ such that the sum $\sum_{k=1}^\infty x_k$ converges, then the sum $\sum_{k=1}^\infty \|x_k\|^2$ also converges and

$$\left\| \sum_{k=1}^\infty x_k \right\|^2 = \sum_{k=1}^\infty \|x_k\|^2.$$

**7.22.** (a) Let $E_1$ and $E_2$ be subspaces of an inner product space. Prove that $E_1 \perp E_2$ if and only if

$$\|x_1 + x_2\|^2 = \|x_1\|^2 + \|x_2\|^2$$

whenever $x_1 \in E_1$, $x_2 \in E_2$.
(b) In contrast to part (a), give an example of a Hilbert space $H$ and vectors $x_1, x_2 \in H$ such that $\|x_1 + x_2\|^2 = \|x_1\|^2 + \|x_2\|^2$, but $\langle x_1, x_2 \rangle \neq 0$.

**7.23.** Let $E$ be a closed convex subset of a Hilbert space. Prove that for every $x \in H$, there is a unique $y \in E$ such that

$$\|x - y\| = d(x, E) = \inf_{u \in E} \|x - u\|.$$

Show that the statement does not necessarily hold if $E$ is not closed or not convex.

**7.24.** Prove that the family

$$\left\{ \frac{1}{\sqrt{\pi}} \sin(nt), \quad \frac{1}{\sqrt{\pi}} \cos(nt), \quad \frac{1}{\sqrt{2\pi}} \right\}_{n \in \mathbf{N}}$$

from Example 7.14 is orthonormal.

**7.25.** Prove Lemma 7.3.

**7.26.** Let $(e_k)$ be an orthonormal sequence in an inner product space $X$. Prove that

$$\sum_{k=1}^{\infty} |\langle x, e_k \rangle \langle y, e_k \rangle| \leq \|x\|\|y\|,$$

for all $x, y \in X$.

**7.27.** Let $\{x_i\}_{i \in J}$ be a summable family of nonnegative numbers with sum $x$. Prove that for every nonempty subset $J_0$ of $J$, one has $\sum_{i \in J_0} x_i \leq x$.

**7.28.** Let $J$ be a countable set and $\{a_i\}_{i \in J}$ a summable family of real or complex numbers with sum $a$. Prove that for every enumeration $J = \{i_1, i_2, \ldots\}$, one has

$$\sum_{k=1}^{\infty} a_{i_k} = a.$$

**7.29.** Show that the family $\{e^{i\lambda t}\}_{\lambda \in \mathbf{R}}$ in Example 7.14 is orthonormal and total in the space $X$.

**7.30.** Let $P$ be the vector space of all real polynomials on $[-1, 1]$. Show that

$$\langle x, y \rangle = \int_{-1}^{1} x(t)y(t)\, dt$$

defines an inner product on $P$. Use the Gram–Schmidt process to orthonormalize the set $\{1, t, t^2\}$.

**7.31.** Show that a countable subset $S$ of a vector space $V$ contains a linearly independent subset $L$ such that $\operatorname{span} L = \operatorname{span} S$.

**7.32.** If $\langle x_1, y \rangle = \langle x_2, y \rangle$ for all $y$ in an inner product space $X$, then $x_1 = x_2$.

**7.33.** Show that (7.17) defines an inner product on $H^*$ that induces a norm on $H^*$.

**7.34.** Let $S$ be a bounded sesquilinear form on $X \times Y$. Show that

$$\|S\| = \sup\left\{ \frac{|S(x,y)|}{\|x\|\|y\|} : x \in X \setminus \{0\},\ y \in Y \setminus \{0\} \right\}$$

and

$$|S(x,y)| \leq \|S\|\|x\|\|y\|,$$

for all $x \in X$ and $y \in Y$.

**7.35.** Let $X$ and $Y$ be inner product spaces and $T : X \to Y$ a bounded linear operator. Show that $T = 0$ if and only if $\langle Tx, y \rangle = 0$ for all $x \in X$ and $y \in Y$.

**7.36.** Let $X$ be a complex inner product space and $T : X \to X$ a bounded linear operator. Show that $\langle Tx, x \rangle = 0$ for all $x \in X$ implies $T = 0$.

Show that the conclusion may not hold if $X$ is real. (Hint: Consider rotations in the real plane $\mathbf{R}^2$.)

**7.37.** Let $T : \ell_2 \to \ell_2$ be defined by

$$T : (x_1, \ldots, x_n, \ldots) \mapsto (x_1, \ldots, \tfrac{1}{n}x_n, \ldots).$$

Show that $\mathcal{R}(T)$ is not closed in $\ell_2$.

**7.38.** Show that $\overline{\mathcal{R}(T^*)} = \mathcal{N}(T)^\perp$ and $\overline{\mathcal{R}(T)} = \mathcal{N}(T^*)^\perp$ (cf. Theorem 7.23).

**7.39.** Let $S = I + T^*T$, where $T : H \to H$ is a bounded operator. Show that $S$ is a one-to-one mapping from $H$ to $S(H)$.

**7.40.** Let $T$ be a bijective self-adjoint operator on a Hilbert space. Show that $T^{-1}$ is self-adjoint.

**7.41.** Verify the statements in Example 7.19.

**7.42.** Show that $T^* - \overline{\lambda}I$ is the adjoint operator of $T - \lambda I$, where $T$ is a bounded operator on a Hilbert space and $\lambda \in \mathbf{F}$.

**7.43.** Show that the operator $T - \lambda I$, where $T$ is a normal operator on a Hilbert space, is also normal (cf. Exercise 7.42).

**7.44.** Show that the right shift operator on $\ell_2$ has no eigenvalues, whereas every $\lambda$ such that $|\lambda| < 1$ is an eigenvalue of the left shift operator on $\ell_2$ (cf. Example 7.16).

**7.45.** Show that if $T$ is a self-adjoint operator, then so is $T^n$, where $n \in \mathbf{N}$.

**7.46.** Let $T$ be a bounded operator on a complex Hilbert space $H$.

(a) Show that the operators

$$T_1 = \frac{1}{2}(T + T^*) \qquad \text{and} \qquad T_1 = \frac{1}{2i}(T - T^*)$$

are self-adjoint.
(b) Show that $T$ is normal if and only if the operators $T_1$ and $T_2$ commute.

**7.47.** Prove that if $T : H \to H$ is a self-adjoint operator and $T \neq 0$, then $T^n \neq 0$ for all $n \in \mathbf{N}$.

**7.48.** Show that an isometric operator $T : H \to H$ that is not unitary maps the Hilbert space $H$ onto a proper closed subspace of $H$.

**7.49.** Let $X$ be a finite-dimensional inner product space and $T : X \to X$ an isometric linear operator. Show that $T$ is unitary.

**7.50.** Suppose that a sequence $(T_n)$ of normal operators on a Hilbert space $H$ converges (in operator norm) to an operator $T$. Show that $T$ is a normal operator.

**7.51.** Suppose that $S$ and $T$ are normal operators such that $ST^* = T^*S$. Show that $S + T$ and $ST$ are normal operators.

**7.52.** Show that if $P$ is a projection operator, then $I - P$ is also a projection operator. Moreover, $\|I - P\| = 1$ if $P \neq I$.

**7.53.** If $P$ is a projection operator on a Hilbert space $H$, then

$$\|Px - Py\| \leq \|x - y\|,$$

for all $x, y \in H$, that is, $P$ is a nonexpansive operator.

**7.54.** Show that two closed subspaces $E$ and $E'$ of a Hilbert space are orthogonal if and only if $P_E P_{E'} = 0$.

**7.55.** If $P$ and $Q$ are projections of a Hilbert space $H$ onto closed subspaces $E$ and $E'$, respectively, and $PQ = QP$, then

$$P + Q - PQ$$

is a projection of $H$ onto $E + E'$.

**7.56.** Prove that $P_E P_{E'} = P_E$ if and only if $E \subseteq E'$.

**7.57.** Show that $U$ is a self-adjoint unitary operator if and only if $U = 2P - I$ for some projection operator $P$.

**7.58.** Show that $U$ is a self-adjoint unitary operator on a Hilbert space $H$ if and only if there exist orthogonal closed subspaces $E_1$, $E_2$ such that $H$ is the direct sum $E_1 \oplus E_2$ and for every $x = x_1 + x_2$ with $x_1 \in E_1$, $x_2 \in E_2$,

$$Ux = x_1 - x_2,$$

that is, $U$ is a reflection.

# Appendix A
# The Hilbert Spaces $\ell_2(J)$

In this appendix, we describe in detail the Hilbert spaces $\ell_2(J)$ that were introduced in passing in Example 7.13. For this we need more results on summable families than those found in Section 7.3. This material is covered in Section A.1, while the Hilbert spaces are the subject of Section A.2.

## A.1 Summable Families of Vectors in a Normed Space

In what follows, $J$ is a nonempty set of indices (finite or infinite) and $\{x_i\}_{i \in J}$ a family of vectors in a normed space $X$. For a finite subset $I \subseteq J$, $x_I$ stands for the *partial sum* $\sum_{k \in I} x_k$. We begin by expanding Definition 7.6 from Section 7.3.

**Definition A.1.** A family $\{x_i\}_{i \in J}$ in $X$ is said to be *summable with sum* $x \in X$ if for every $\varepsilon > 0$, there exists a finite subset $J_0$ of $J$ such that

$$\|x - x_I\| < \varepsilon$$

for every finite subset $I$ of $J$ containing $J_0$. If $\{x_i\}_{i \in J}$ is summable with sum $x$, we write $\sum_{i \in J} x_i = x$.

A family $\{x_i\}_{i \in J}$ is said to be a *Cauchy family* if for every $\varepsilon > 0$ there exists a finite subset $J_0$ of $J$ such that for every finite subset $K$ disjoint from $J_0$, the inequality $\|x_K\| < \varepsilon$ holds.

**Theorem A.1.** *A family of vectors $\{x_i\}_{i \in J}$ in a Banach space $X$ is summable if and only if it is a Cauchy family.*

*Proof.* (Necessity.) Suppose that $\{x_i\}_{i \in J}$ is summable with sum $x$. For $\varepsilon > 0$, there exists a finite set $J_0 \subseteq J$ such that if a finite set $I$ contains $J_0$, then $\|x_I - x\| < \varepsilon/2$. Let $K$ be a finite subset of $J$ disjoint from $J_0$. Then

$$\|x_{J_0} - x\| < \frac{\varepsilon}{2} \qquad \text{and} \qquad \|x_{J_0 \cup K} - x\| < \frac{\varepsilon}{2}.$$

© Springer International Publishing AG, part of Springer Nature 2018
S. Ovchinnikov, *Functional Analysis*, Universitext,
https://doi.org/10.1007/978-3-319-91512-8

Hence

$$\|x_K\| = \|x_{J_0 \cup K} - x_{J_0}\| = \|x_{J_0 \cup K} - x + x - x_{J_0}\|$$
$$\leq \|x_{J_0 \cup K} - x\| + \|x_{J_0} - x\| < \varepsilon.$$

It follows that $\{x_i\}_{i \in J}$ is a Cauchy family.

(Sufficiency.) Suppose that $\{x_i\}_{i \in J}$ is a Cauchy family and let $(\varepsilon_k)$ be a strictly decreasing sequence of positive numbers converging to zero. We construct inductively a nested sequence of closed subsets of $X$ whose intersection is a singleton $\{x\}$ in $X$ and show that $\{x_i\}_{i \in J}$ is summable with sum $x$.

Step 1. Let $\varepsilon = \varepsilon_1$. Inasmuch as the family $\{x_i\}_{i \in J}$ is Cauchy, there exists a finite subset $J_1$ such that for every finite set $K$ disjoint from $J_1$, we have $\|x_K\| < \varepsilon_1/2$. Therefore, for every two finite sets $I'$ and $I''$ containing $J_1$, we have

$$\|x_{I'} - x_{J_1}\| = \|x_{I' \setminus J_1}\| < \frac{\varepsilon_1}{2} \quad \text{and} \quad \|x_{I''} - x_{J_1}\| = \|x_{I'' \setminus J_1}\| < \frac{\varepsilon_1}{2}.$$

It follows that

$$\|x_{I'} - x_{I''}\| < \varepsilon_1.$$

We define

$$X_1 = \{x_I : J_1 \subseteq I, \ I \text{ a finite set}\}.$$

Clearly, $\operatorname{diam}(X_1) \leq \varepsilon_1$.

Step 2. Let $\varepsilon = \varepsilon_2$. By the same arguments as in Step 1, there is a finite subset $J_2$ of $J$ such that for every $I'$ and $I''$ containing $J_2$, we have

$$\|x_{I'} - x_{I''}\| < \varepsilon_2.$$

We define

$$X_2 = \{x_I : J_1 \cup J_2 \subseteq I, \ I \text{ a finite set}\}.$$

Clearly, $X_2 \subseteq X_1$ and $\operatorname{diam}(X_2) \leq \varepsilon_2$.

By continuing this process inductively, we obtain sequences of sets $(J_n)$ and $(X_n)$ such that

$$X_n = \left\{x_I : I \supseteq \bigcup_{k=1}^{n} J_k\right\} \quad \text{and} \quad \operatorname{diam}(X_n) \leq \varepsilon_n.$$

It is clear from the inductive construction that $X_{n+1} \subseteq X_n$, which implies $\overline{X}_{n+1} \subseteq \overline{X}_n$.

Because $\operatorname{diam}(\overline{X}_n) = \operatorname{diam}(X_n)$ (cf. Exercise 2.39) and $\varepsilon_n \to 0$, we conclude (cf. Theorem 2.9) that the intersection $\bigcap_{k=1}^{\infty} X_k$ is a singleton, say $\{x\}$. We show that $x$ is the sum of the family $\{x_i\}_{i \in J}$.

For a given $\varepsilon > 0$, we choose $n$ such that $\varepsilon_n < \varepsilon$. For $J_0 = \bigcup_{k=1}^{n} J_k$ and every finite $I \supseteq J_0$, we have $x_I \in X_n$. Since $x \in \overline{X}_n$ and $\operatorname{diam}(\overline{X}_n) \leq \varepsilon_n < \varepsilon$, it follows that $\|x_I - x\| < \varepsilon$. Therefore, $\sum_{i \in J} x_i = x$.                $\square$

Two properties of summable families are the subject of the next theorem.

**Theorem A.2.** *Let $\{x_i\}_{i \in J}$ and $\{y_i\}_{i \in J}$ be summable families in a normed space $X$ with sums $x$ and $y$, respectively. Then:*

(a) *The family $\{x_i + y_i\}_{i \in J}$ is summable with sum $x + y$.*
(b) *For every $\lambda \in \mathbf{F}$, the family $\{\lambda x_i\}_{i \in J}$ is summable with sum $\lambda x$.*

*Proof.* (a) Let $\varepsilon > 0$. There exist finite subsets $J_0'$ and $J_0''$ of the set $J$ such that
$$\|x - x_I\| < \frac{\varepsilon}{2} \quad \text{and} \quad \|y - y_I\| < \frac{\varepsilon}{2},$$
for every finite set $I$ containing $J_0 = J_0' \cup J_0''$. We have
$$\|x + y - (x + y)_I\| = \|x + y - x_I - y_I\| \le \|x - x_I\| + \|y - y_I\| < \varepsilon.$$

Because $\varepsilon$ is an arbitrary positive number, the family $\{x_i + y_i\}_{i \in J}$ is summable.

(b) The claim is trivial for $\lambda = 0$, so we may assume that $\lambda \ne 0$. For every $\varepsilon > 0$, there exists a finite set $J_0 \subseteq J$ such that
$$\|x - x_I\| < \frac{\varepsilon}{|\lambda|}$$
for every finite set $I$ containing $J_0$. Hence, $\|\lambda x - (\lambda x)_I\| < \varepsilon$, and the result follows. $\qquad\square$

Another property of a summable family is the result of the next theorem.

**Theorem A.3.** *Let $\{x_i\}_{i \in J}$ be a summable family in a normed space $X$. For every $\varepsilon > 0$ there exists a finite set $J_0$ such that $\|x_i\| < \varepsilon$ for all $i \in J \setminus J_0$.*

*Proof.* Let $x = \sum_{i \in J} x_i$. There exists a finite set $J_0 \subseteq J$ such that for every finite set $I$ containing $J_0$,
$$\|x - x_I\| < \frac{\varepsilon}{2}.$$
Let $i \in J \setminus J_0$ and $I = J_0 \cup \{i\}$. We have
$$\|x - x_I\| < \frac{\varepsilon}{2} \quad \text{and} \quad \|x - x_{J_0}\| < \frac{\varepsilon}{2}.$$
Then
$$\|x_i\| = \|x_I - x_{J_0}\| = \|(x - x_{J_0}) - (x - x_I)\| < \varepsilon,$$
because $\{i\} = I \setminus J_0$. $\qquad\square$

**Corollary A.1.** (Cf. Lemma 7.4.) *There are at most countably many nonzero vectors in a summable family.*

*Proof.* Let $\{x_i\}_{i \in J}$ be a summable family in $X$, $J' = \{i \in J : x_i \neq 0\}$, and

$$J_n = \{i \in J : \|x_i\| > 1/n\}.$$

By Theorem A.3, $J_n$ is a finite set for every $n \in \mathbf{N}$. Inasmuch as $J' = \cup_{n \in \mathbf{N}} J_n$, the set $J'$ is at most countable. □

In spite of obvious similarities between convergent series and summable families of numbers, these are quite different notions, as the following example suggests.

*Example A.1.* As is well known, the alternating series

$$\sum_{k=1}^{\infty} \frac{(-1)^{k+1}}{k} = 1 - \frac{1}{2} + \frac{1}{3} - \frac{1}{4} + \cdots$$

converges. However, the family $\{x_k = \frac{(-1)^{k+1}}{k}\}_{k \in \mathbf{N}}$ is not summable. Indeed, otherwise, for $\varepsilon = 1$, there must be a finite set $J_0 \subseteq \mathbf{N}$ such that the set of partial sums

$$\{x_I : I \text{ is a finite subset of } \mathbf{N} \text{ containing } J_0\}$$

is bounded. For if $x = \sum_{i \in \mathbf{N}} x_i$, then $\|x - x_I\| < 1$ for all finite sets $I \subseteq \mathbf{N}$ containing $J_0$. Let $N = \max J_0$. We may assume that $N$ is an even number. Let

$$I_n = J_0 \cup \{1/(N+1)\} \cup \cdots \cup \{1/(N+2n-1)\}$$

for $n \in \mathbf{N}$. Because the harmonic series diverges, the family of partial sums $\{x_{I_n}\}$ is unbounded. This contradicts our assumption that the family $\{x_i\}_{i \in \mathbf{N}}$ is summable.

The last theorem in this section is an analogue of the theorem in analysis claiming that an absolutely convergent series converges.

**Definition A.2.** A family $\{x_i\}_{i \in J}$ in $X$ is said to be *absolutely summable* if the family of real numbers $\{\|x_i\|\}_{i \in J}$ is summable.

**Theorem A.4.** *If a family of vectors $\{x_i\}_{i \in J}$ in a Banach space $X$ is absolutely summable, then it is summable.*

*Proof.* Let $\varepsilon > 0$. Because $\{\|x_i\|\}_{i \in J}$ is summable, by Theorem A.1, there exists a finite subset $J_0$ of the set $J$ such that for every finite set $K \subseteq J$ disjoint from $J_0$, we have $\sum_{i \in K} \|x_i\| < \varepsilon$. Then

$$\left\| \sum_{i \in K} x_i \right\| \leq \sum_{i \in K} \|x_i\| < \varepsilon.$$

By Theorem A.1, the result follows. □

## A.2 The Hilbert Spaces $\ell_2(J)$

Let $J$ be a nonempty set. We denote by $\ell_2(J)$ the set of all families $\{x_i\}_{i \in J}$ of numbers in $\mathbf{F}$ such that the family $\{|x_i|^2\}_{i \in J}$ is summable. Operations of addition and scalar multiplication on $\ell_2(J)$ are defined by

$$\{x_i\}_{i \in J} + \{y_i\}_{i \in J} = \{x_i + y_i\}_{i \in J} \quad \text{and} \quad \lambda\{x_i\}_{i \in J} = \{\lambda x_i\}_{i \in J}, \qquad (A.1)$$

respectively.

**Theorem A.5.** *The set $\ell_2(J)$ is a vector space with respect to the operations of addition and scalar multiplication defined by (A.1).*

*Proof.* Suppose that $\{x_i\}_{i \in J}$ and $\{y_i\}_{i \in J}$ are in $\ell_2(J)$. For every $i \in J$, we have

$$|x_i + y_i|^2 \le 2(|x_i|^2 + |y_i|^2)$$

(cf. Exercise 1.5 (c)). By Theorem A.2, the family $\{2(|x_i|^2 + |y_i|^2)\}_{i \in J}$ is summable. By Lemma 7.3, the family $\{|x_i + y_i|^2\}_{i \in J}$ is summable. Hence $\{x_i + y_i\}_{i \in J} \in \ell_2(J)$, so $\ell_2(J)$ is closed under addition.

It is clear that $\{\lambda x_i\}_{i \in J} \in \ell_2(J)$ for every $\lambda \in \mathbf{F}$, provided that $\{x_i\}_{i \in J}$ is in $\ell_2(J)$. Thus $\ell_2(J)$ is closed under multiplication by scalars. □

Let $x = \{x_i\}_{i \in J}$ and $y = \{y_i\}_{i \in J}$ be vectors in $\ell_2(J)$. The inner product on the space $\ell_2(J)$ is defined by

$$\langle x, y \rangle = \sum_{i \in J} x_i \overline{y_i}. \qquad (A.2)$$

The next theorem justifies this definition.

**Theorem A.6.** *For every two families $\{x_i\}_{i \in J}$ and $\{y_i\}_{i \in J}$ in $\ell_2(J)$, the family $\{x_i \overline{y_i}\}_{i \in J}$ is summable.*

*Proof.* For every $i \in J$,

$$2|x_i||y_i| \le |x_i|^2 + |y_i|^2.$$

By Theorem A.2, the family $\{|x_i|^2 + |y_i|^2\}_{i \in J}$ is summable. By Lemma 7.3, the family $\{|x_i y_i|\}_{i \in J}$ is summable. Hence the family $\{x_i \overline{y_i}\}_{i \in J}$ is absolutely summable. By Theorem A.4, this family is summable. □

It is not difficult to verify that $\langle x, y \rangle$ defined by (A.2) is indeed an inner product on $\ell_2(J)$, so $\ell_2(J)$ is an inner product space. The norm induced by the inner product on $\ell_2(J)$ is given by

$$\|x\| = \sqrt{\langle x, x \rangle} = \sqrt{\sum_{i \in J} |x_i|^2}.$$

The inner product space $\ell_2(J)$ is a Hilbert space, as the next theorem claims. In the proof, we imitate the steps from the proof of Theorem 3.5.

**Theorem A.7.** *The inner product space $\ell_2(J)$ is complete.*

*Proof.* Let $(x^{(n)})$ be a Cauchy sequence in $\ell_2(J)$, where $x^{(n)} = \{x_i^{(n)}\}_{i \in J}$. Then for every $\varepsilon > 0$, there exists $N \in \mathbf{N}$ such that for all $m, n > N$,

$$\|x^{(m)} - x^{(n)}\| = \sqrt{\sum_{i \in J} |x_i^{(m)} - x_i^{(n)}|^2} < \varepsilon. \qquad (A.3)$$

Therefore, for every $i \in J$,

$$|x_i^{(m)} - x_i^{(n)}| < \varepsilon, \qquad \text{for all } m, n > N$$

(cf. Exercise 7.27). It follows that for every $i \in J$, the sequence $(x_i^{(n)})$ is a Cauchy sequence in $\mathbf{F}$ and therefore converges to some limit $x_i$ as $n \to \infty$. We define $x = \{x_i\}_{i \in J}$ and show that $x \in \ell_2(J)$ and $x^{(m)} \to x$ as $m \to \infty$.

Let $I$ be a finite subset of $J$. By the result of Exercise 7.27, we obtain from (A.3)

$$\sum_{i \in I} |x_i^{(m)} - x_i^{(n)}|^2 < \varepsilon^2, \qquad \text{for all } m, n > N.$$

By taking the limit as $n \to \infty$ (note that $I$ is a finite set), we obtain

$$\sum_{i \in I} |x_i^{(m)} - x_i|^2 \leq \varepsilon^2, \qquad \text{for all } m > N.$$

By Lemma 7.2, we have

$$\sum_{i \in J} |x_i^{(m)} - x_i|^2 \leq \varepsilon^2, \qquad \text{for all } m > N. \qquad (A.4)$$

It follows that $(x - x^{(m)}) \in \ell_2(J)$. Because $x = x^{(m)} + (x - x^{(m)})$, we have proved that $x \in \ell_2(J)$.

Finally, by (A.4), $\|x^{(m)} - x\| \leq \varepsilon$ for all $m > N$. Hence $x^{(m)} \to x$ in $\ell_2(J)$. The result follows. $\qquad \square$

For $k \in J$, we define a vector $e_k \in \ell_2(J)$ as a family $\{x_i\}_{i \in J}$ such that

$$x_i = \begin{cases} 1, & \text{if } i = k, \\ 0, & \text{if } i \neq k. \end{cases}$$

It is easy to verify that the family $\{e_k\}_{k \in J}$ is an orthonormal family in $\ell_2(J)$. Because for $x = \{x_i\}_{i \in J}$, we have $\langle x, e_k \rangle = x_k$, the zero vector is the only vector in $\ell_2(J)$ that is orthogonal to the family $\{e_k\}_{k \in J}$. Hence $\{e_k\}_{k \in J}$ is an orthonormal basis of $\ell_2(J)$.

## Notes

Our coverage of summable families in normed spaces and Hilbert spaces $\ell_2(J)$ is based on presentations of these subjects in Dixmier (1984), Bourbaki (1966), and Bourbaki (2003).

As we mentioned before, there is a considerable difference between the notions of convergent series and summable families (cf. Example A.1). The following statement is another illustration of this difference: A family of real or complex numbers is summable if and only if it is absolutely summable (Theorem 9.4.6 in Dixmier 1984).

The spaces $\ell_2^n$ and $\ell_2$ are instances of the spaces $\ell_2(J)$ for $J = \{1, \ldots, n\}$ and $J = \mathbf{N}$, respectively.

Every Hilbert space is isomorphic to a space $\ell_2(J)$ (Theorem 9.5.5 in Dixmier 1984).

# References

Axler, S. (2015). *Linear Algebra Done Right*. Springer, 3rd edition.

Banach, S. (1987). *Theory of Linear Operators*. North-Holland, Amsterdam.

Bourbaki, N. (1966). *General Topology*. Addison-Wesley Publishing Company.

Bourbaki, N. (2003). *Topological Vector Spaces*. Springer-Verlag, Berlin, Heidelberg, New York.

Dixmier, J. (1984). *General Topolgy*. Springer-Verlag, New York, Berlin, Heidelberg, Tokyo.

Dunford, N. and Schwartz, J. T. (1958). *Linear Operators, Part I*. Interscience Publishers, New York, London.

Halmos, P. R. (1960). *Naive Set Theory*. Springer.

Halmos, P. R. (1967). *A Hilbert Space Problem Book*. D. Van Nostrand Company, Inc., Princeton, New Jersey, Toronto, London.

Kiselev, A. P. (2006). *Kiselev's Geometry, Book I, Planimetry*. Sumizdat, El Cerrito, CA.

Lindenstrauss, J. and Tzafriri, L. (1977). *Classical Banach Spaces I: Sequence Spaces*. Springer-Verlag, Berlin, Heidelberg, New York.

Munkres, J. R. (2000). *Topology*. Prentice Hall, Upper Saddle River, NJ 07458, 2nd edition.

Ovchinnikov, S. (2015). *Number Systems*. American Mathematical Society, Providence, Rhode Island.

Royden, H. L. and Fitzpatrick, P. M. (2010). *Real Analysis*. Prentice Hall, Upper Saddle River, NJ 07458, 4th edition.

Rudin, W. (1973). *Functional Analysis*. McGraw-Hill, Inc.

Wade, W. R. (2000). *An Introduction to Analysis*. Prentice Hall, Upper Saddle River, NJ 07458, 2nd edition.

© Springer International Publishing AG, part of Springer Nature 2018
S. Ovchinnikov, *Functional Analysis*, Universitext,
https://doi.org/10.1007/978-3-319-91512-8

# Index

© Springer International Publishing AG, part of Springer Nature 2018
S. Ovchinnikov, *Functional Analysis*, Universitext,
https://doi.org/10.1007/978-3-319-91512-8

Printed in the United States
By Bookmasters